An Illustrated Atlas of the
Skeletal Muscles
THIRD EDITION

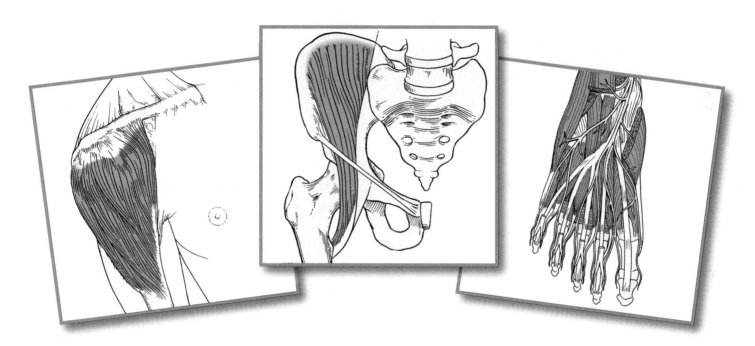

Bradley S. Bowden
Alfred University

Joan M. Bowden
Alfred University

Principal Illustrator, 3rd edition
Peggy Firth

Morton Publishing Company
925 W. Kenyon Avenue, Unit 12
Englewood, CO 80110
800-348-3777
www.morton-pub.com

Book Team

Publisher	:	Douglas Morton
Editor/Project Coordinator	:	Dona Mendoza
Interior Design/Production	:	Joanne Saliger, Ash Street Typecrafters, Inc.
Cover Design	:	Bob Schram, Bookends

Library of Congress Control Number: 2009926156

ISBN: 978-089582-808-8

Printed in the United States of America

10   9   8   7   6   5   4   3   2   1

Preface

An *Illustrated Atlas of the Skeletal Muscles* is a functional reference and study guide for students and health professionals. In response to excellent suggestions from adopters and reviewers, several useful modifications and additions have been included in the Third Edition. The unique "information boxes" containing brief descriptions of representative problems in both the skeleton and muscles resulting from injury and disease have been greatly expanded. Travell & Simons' *Trigger Point Manual* was the reference for the trigger points included in the "information boxes" in the chapters that illustrate the muscles. Trigger points are signified by a bullet (•). Names of bones, muscles, and nerves are accompanied by phonetic spellings (*Merriam Webster's Medical Desk Dictionary*) to assist in pronunciation.

The first chapter presents different views of the human skeleton, with bones and bone groups clearly labeled. "Bony features" that define either articulating surfaces or designate specific points of muscle attachment are also indicated. "Bony features" and muscles are labeled by numbers that correspond to the numbers in the key on each page. This enables the user to cover the key and use the diagrams as a quiz. New palpation descriptions with the bony landmark illustrations enable the user to locate major bony features on themselves or on a partner. Several new illustrations of primary types of bone fractures are followed by a series of seven radiographs, line drawings, and descriptions of prevalent youth sports growth plate fractures.

In Chapter 2, illustrations of the primary types of articulations between bones enable the user to thoroughly understand the degree and range of motion resulting from muscle contraction. New wrist-hand and ankle-foot ligament illustrations complete the pectoral and pelvic appendage ligament series.

Chapter 3 illustrates the diversity of movements in the body. Inclusion of the articulating skeletal elements reinforces the names and locations of bones, their association in articulations, the specific sites of muscle attachment and the muscle movements permitted. A special feature is a three-page set of illustrations from the March 2001 issue of *Scientific American* depicting skeleto-muscular "flaws" in the engineering of the human body that may be the bases of many of the human skeleto-muscular problems resulting from sports and the aging process. Also included are some thought-provoking "fixes" for these conditions.

A second special feature is an eight-page, full-color insert illustrating the human systems. It includes the skeletal, muscular, circulatory, respiratory, digestive, urinary and endocrine, and nervous systems.

Several changes are seen in muscle Chapters 4 through 10. Most significant is the addition of 48 new illustrations arranged in 38 functional muscle groups, providing students with the opportunity to understand the combined action of multiple muscles in producing primary body movements. As previously, each muscle is presented in color with specific points of skeletal attachment clearly indicated. Each muscle is given both Latin and anglicized names following *Anatomica Terminologica*. The origin and insertion (with alternate attachment terms), muscle actions, nerve innervation, and palpation descriptions are listed with each muscle. The accompanying information box includes practical information on muscle structural and attachment variations, developmental anomalies, injuries, trigger point location, referred pain pattern and the muscle's synergists and antagonists.

The *Atlas* also includes illustrations of the muscles of the eye and tympanic cavity, along with illustrations of the tongue, larynx, pharynx and palate.

Finally, the series of illustrations of innervation pathways for major muscles and muscle groups in Chapter 11 is enhanced with the addition of the cervical plexus. These enable the student or professional to appreciate the potential relationship between muscle weakness and damage to specific nerves.

Acknowledgments

We would like to thank Douglas Morton and Dona Mendoza for initially proposing this project and convincing us of the need for a skeleto-muscular atlas of this type. As Project Manager, Dona Mendoza kept everyone on task and on target, attending to the innumerable details and coordinating work between the authors and the illustrators. Peggy Firth and her staff at Firth Studies were responsible for the majority of the new illustrations; Winnie Luong and the staff at Imagineering Media Services provided the youth sports injury illustrations. The radiograph images were made possible through the generosity and permission of the following:

Dr. Adam Bird, Department of Podiatry, LaTrobe University, Bundoora, Victoria, Australia

William Herring, M.D., Department of Radiology, Albert Einstein Medical Center, Philadelphia, PA. LearningRadiology.com

Brian Tidey, Tauranga, New Zealand, www.radiographersreproting.com

We also wish to express our gratitude to various adopters for invaluable comments, suggestions, and criticisms, and especially to Chloe Costigan-Hume from Quinnipiac University for her extensive analysis and suggestions on terminology.

We thank artist Patricia J. Wynne, www.patriciawynne.com, for non-exclusive printing rights for her illustrations and to the *Scientific American, Inc.* for the article, "If Humans Were Built to Last," by S. Jay Olshanky, Bruce A. Carnes, and Robert N. Butler. Copyright © 2001 by Scientific American, Inc. All rights reserved.

The Authors

Contents

5 Muscles of the Neck 99

An Illustrated Atlas of the Skeletal Muscles

8 Muscles of the Forearm and Hand 181

9 Muscles of the Hip and Thigh 211

10 Muscles of the Lower Leg and Foot 237

11 Functional Muscle Groups 263

12 Muscle Innervation Pathways 303

The Skeleton

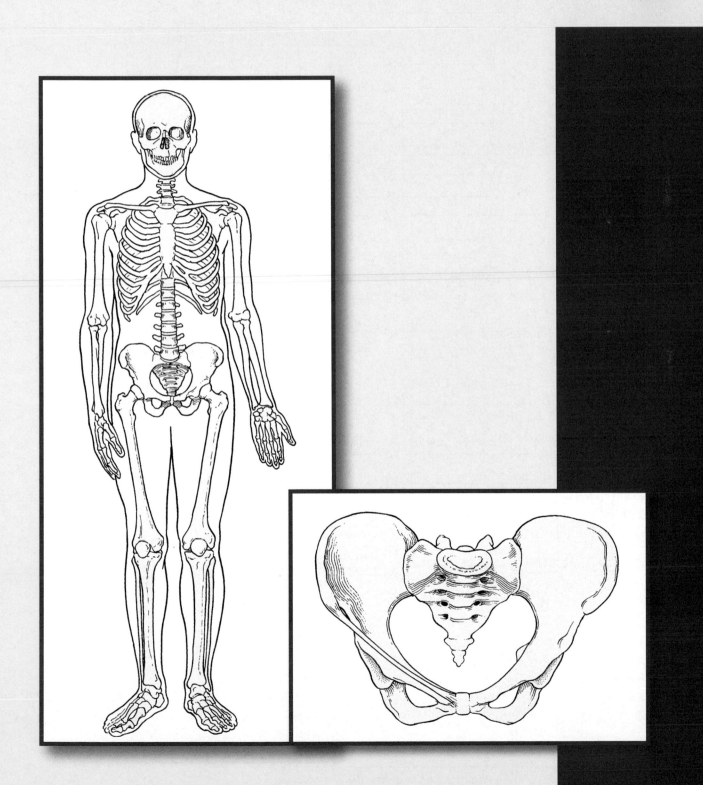

Anterior and Posterior Views of Skeleton

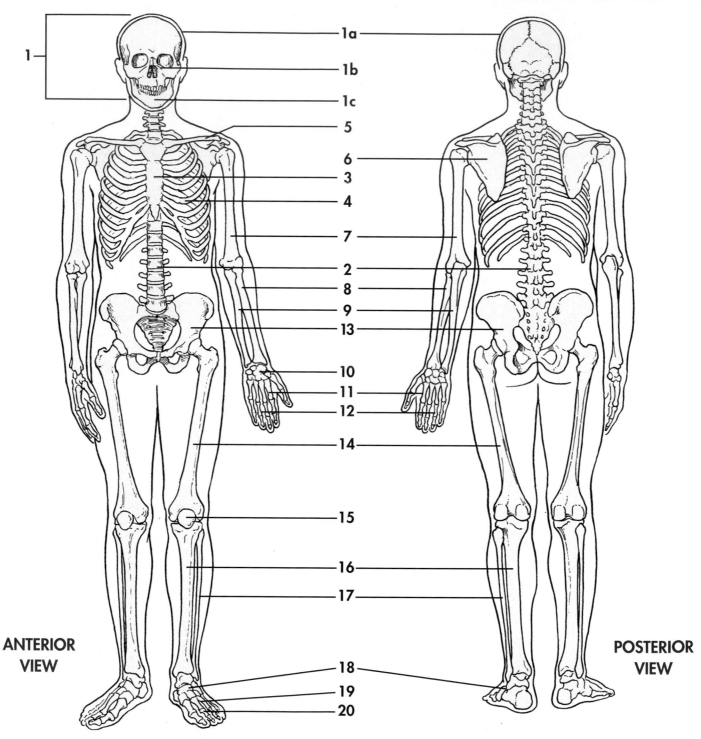

ANTERIOR VIEW

POSTERIOR VIEW

1 Skull	**4** Ribs	**10** Carpals	**16** Tibia
1a Cranium	**5** Clavicle	**11** Metacarpals	**17** Fibula
1b Facial bones	**6** Scapula	**12** Phalanges	**18** Tarsals
1c Mandible	**7** Humerus	**13** Os coxa	**19** Metatarsals
2 Vertebral column	**8** Radius	**14** Femur	**20** Phalanges
3 Sternum	**9** Ulna	**15** Patella	

An Illustrated Atlas of the Skeletal Muscles

Lateral and Anterior Views of Skull

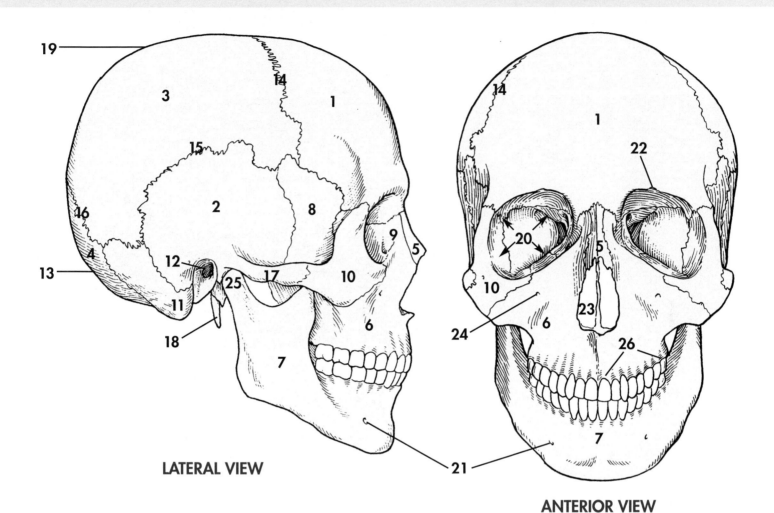

LATERAL VIEW

ANTERIOR VIEW

1	Frontal bone	**10**	Zygomatic bone	**18**	Styloid process of temporal
2	Temporal bone	**11**	Mastoid process of temporal bone	**19**	Sagittal suture
3	Parietal bone	**12**	External auditory meatus	**20**	Bony orbit
4	Occipital bone	**13**	External occipital protuberance	**21**	Mental foramen
5	Nasal bone	**14**	Coronal suture	**22**	Supraorbital foramen
6	Maxilla	**15**	Squamosal suture	**23**	Nasal septum
7	Mandible	**16**	Lambdoidal suture	**24**	Infraorbital foramen
8	Sphenoid bone	**17**	Zygomatic process of temporal	**25**	Mandibular condyle
9	Lacrimal bone			**26**	Maxillary alveolar process

The **temporomandibular joint** usually dislocates anteriorly. During yawning or taking a large bite, contraction of the **lateral pterygoid** muscles may cause the heads of the mandible to dislocate. In this position, a person is unable to close his mouth. Dislocation can also occur during tooth extraction. Most commonly, the dislocation is caused by a blow to the chin when the mouth is open. Fractures of the maxillary bone (upper jaw or midface) are referred to as **Leforte fractures**.

Superior and Inferior Views of Skull

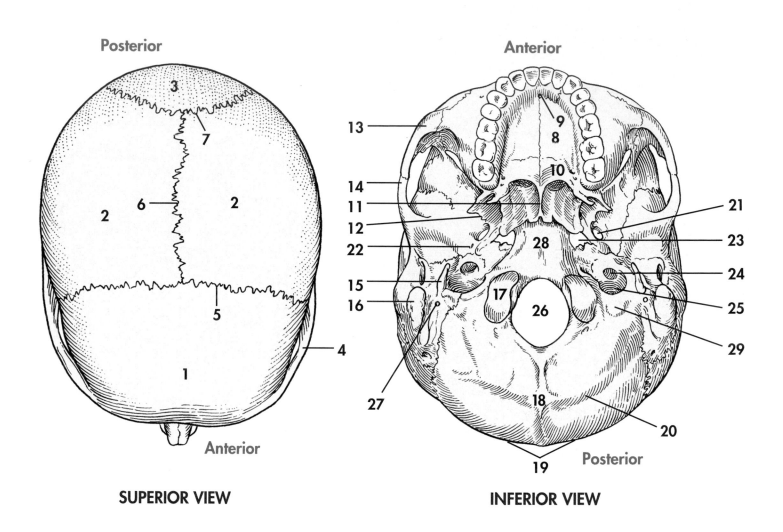

Posterior

7

3

6   2

2

5

4

1

Anterior

SUPERIOR VIEW

Anterior

13

9

8

14

10

11

12   21

22   28   23

15   24

16   17   25

26   29

18

27   20

19   Posterior

INFERIOR VIEW

1 Frontal bone	**11** Vomer bone	**21** Foramen ovale
2 Parietal bone	**12** Sphenoid bone	**22** Foramen spinosum
3 Occipital bone	**13** Zygomatic bone	**23** Foramen lacerum
4 Zygomatic arch	**14** Zygomatic process of temporal	**24** Carotid canal
5 Coronal suture	**15** Styloid process of temporal	**25** Jugular foramen
6 Sagittal suture	**16** Mastoid process of temporal	**26** Foramen magnum
7 Lambdoidal suture	**17** Occipital condyle	**27** Stylomastoid foramen
8 Palatine process of maxilla	**18** Occipital protuberance	**28** Basilar process of occipital
9 Incisive foramen	**19** Superior nuchal line	**29** Jugular process
10 Palatine bone	**20** Inferior nuchal line	

Internal View of Base of Skull

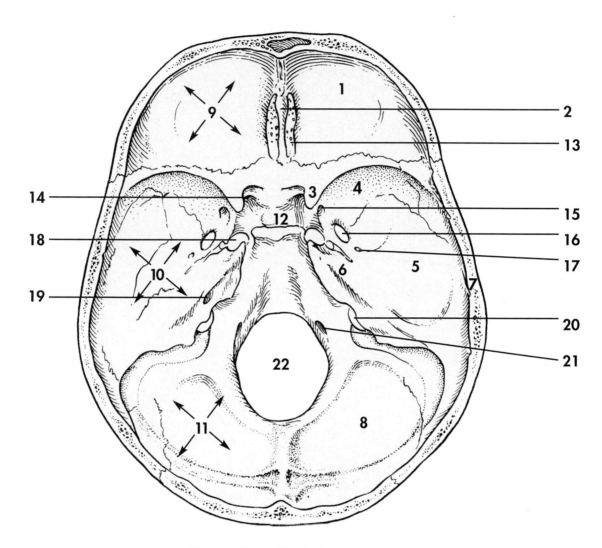

SUPERIOR INTERNAL VIEW

1 Frontal bone
2 Crista galli of ethmoid bone
3 Lesser wing of sphenoid bone
4 Greater wing of sphenoid bone
5 Temporal bone (squamous part)
6 Temporal bone (petrous part)
7 Parietal bone
8 Occipital bone
9 Anterior cranial fossa
10 Middle cranial fossa
11 Posterior cranial fossa

12 Sella turcica of sphenoid bone
13 Cribriform plate of ethmoid bone
14 Optic foramen
15 Foramen rotundum
16 Foramen ovale
17 Foramen spinosum
18 Foramen lacerum
19 Internal acoustic meatus
20 Jugular foramen
21 Hypoglossal canal
22 Foramen magnum

Mandible/Dentary
(**man**•da•ble/**den**•ter•ry)

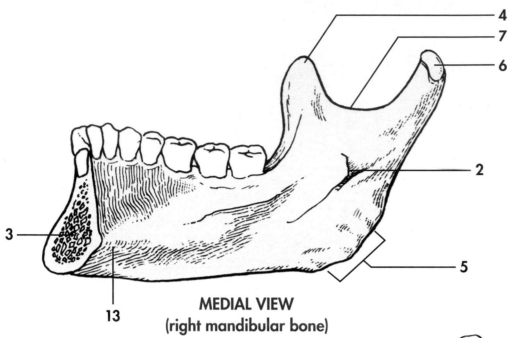

MEDIAL VIEW
(right mandibular bone)

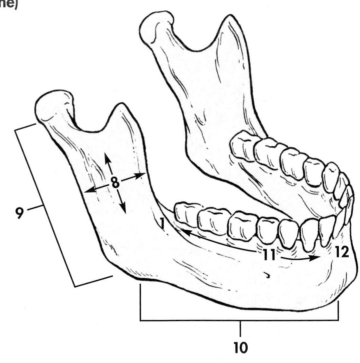

LATERAL INTERNAL VIEW

1 Oblique line
2 Mandibular foramen
3 Mandibular symphysis
4 Coronoid process
5 Mandibular angle
6 Mandibular condyle
7 Mandibular notch
8 Lateral fossa
9 Ramus of mandible
10 Body of mandible
11 Alveolar margin
12 Incisive fossa
13 Inferior mental spine

Cervical Vertebrae, Hyoid Bone, and Thyroid Cartilage
(ser•vi•kal) **(ver**•ta•bray) **(hi**•oid) **(thy**•roid) **(kar**•ti•lege)

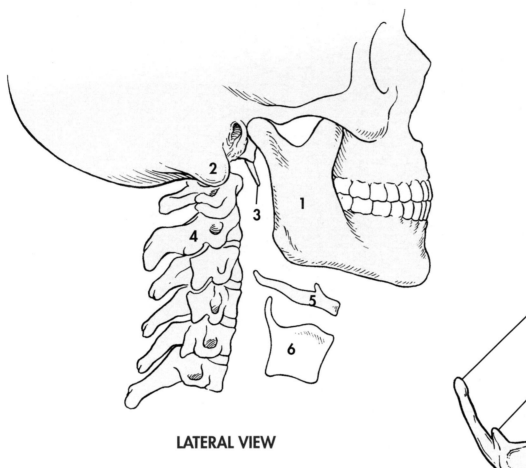

LATERAL VIEW

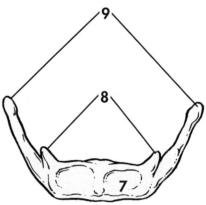

HYOID BONE

ANTERIOR VIEW

1	Mandible	**6**	Thyroid cartilage
2	Mastoid process	**7**	Body of hyoid
3	Styloid process	**8**	Lesser cornu
4	Cervical vertebrae	**9**	Greater cornu
5	Hyoid bone		

The **hyoid bone** does not articulate with any other bone. It is suspended by ligaments from the styloid processes of the temporal bones and serves as a site of attachment for tongue and throat muscles. The hyoid bone is often fractured in incidents of strangulation and is, therefore, carefully examined during an autopsy in which strangulation is suspected. The **thyroid cartilage** is the largest of the nine laryngeal cartilages. The two main plates of this cartilage are fused in front to form a laryngeal prominence (Adam's Apple), which is more pronounced in males and females after puberty. This cartilage can become fractured as a result of blows received during boxing, karate, or compression by a shoulder strap during a vehicle accident. The protective guards hanging from ice hockey goalie masks offer protection against this type of injury.

Sternum and Thoracic Cage
(ster•num)   (tho•ras•ik)

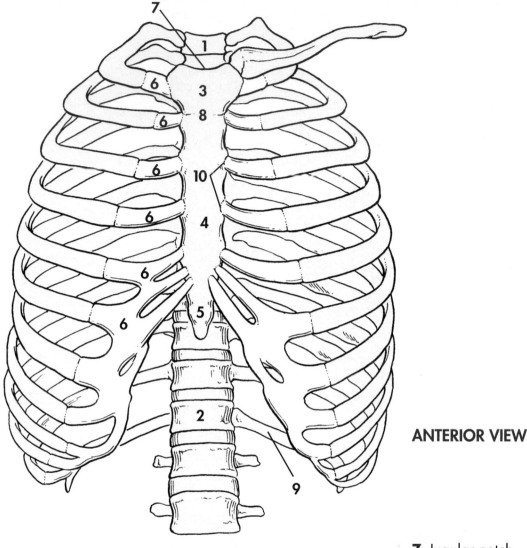

ANTERIOR VIEW

1 First thoracic vertebra	**4** Body of sternum	**7** Jugular notch
2 Twelfth thoracic vertebra	**5** Xiphoid process	**8** Sternal angle
3 Manubrium of sternum	**6** Costal cartilage	**9** Floating rib
		10 Costal notches

Important anatomical landmarks are the **jugular notch, sternal angle,** and **xiphisternal angle.** The **jugular notch** is the midline depression at the top of the **manubrium,** approximately in line with the point from which the left common carotid artery branches from the aorta. The **sternal angle,** a palpable horizontal ridge between the **manubrium** and **body** of the sternum, is in line with the intervertebral disk between the fourth and fifth thoracic vertebrae. It is a reference point for locating the second rib and for listening for heart valve sounds. The **xiphisternal joint** marks the beginning of the xiphoid process. During CPR, excess pressure might break the xiphoid process and puncture the underlying liver or heart. In a **sternal fracture,** the upper portion may displace over or under the lower portion, with potential injury to thoracic organs. During **sternal puncture,** a needle is inserted through the surface into the red marrow to aspirate a sample of red bone marrow for analysis.

Ribs

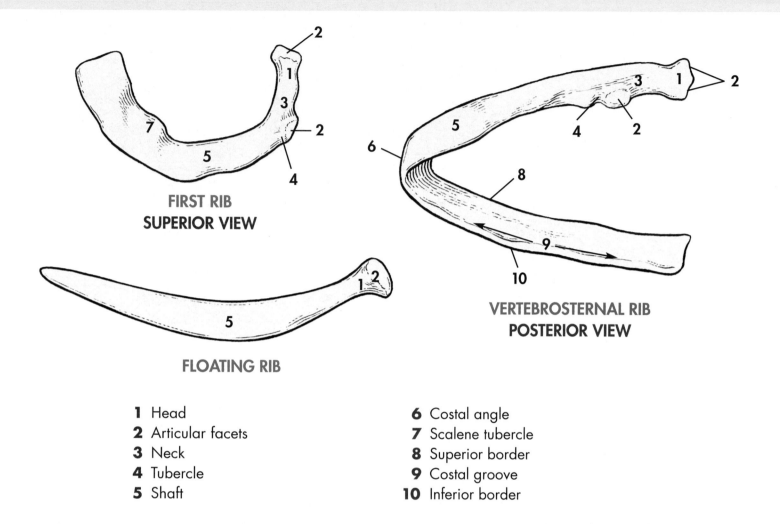

FIRST RIB
SUPERIOR VIEW

FLOATING RIB

VERTEBROSTERNAL RIB
POSTERIOR VIEW

1 Head	**6** Costal angle
2 Articular facets	**7** Scalene tubercle
3 Neck	**8** Superior border
4 Tubercle	**9** Costal groove
5 Shaft	**10** Inferior border

The ribs articulate posteriorly with the thoracic vertebrae and anteriorly (not the last two pair) with the sternum to form the **bony thorax** that houses and protects the organs within, particularly the heart and lungs. The bony thorax, together with the intercostals, diaphragm, and other muscles, provides the skeleto-muscular basis of inspiratory and expiratory respiratory movements. The superior seven pairs of ribs ("**true**" or **vertebrosternal ribs**) attach individually to the sternum by separate **costal cartilages**: the inferior five pairs of **vertebrochondral** or "**false ribs,**" attach to the sternum either jointly (8–10) or not at all (11–12: "**floating ribs**").
A typical rib has a **head, neck, tubercle,** and **shaft.** The wedge-shaped head articulates with two adjacent vertebral bodies by two **facets.** The neck is the short constricted region just lateral to the head. Just lateral to the neck, a knob-like **tubercle** articulates with the transverse process of the vertebra. Next, the shaft angles sharply anterior (the **angle of the rib**) and extends to the costal cartilage (1–10) or ends in the muscle of the abdominal wall (11–12). The shaft's superior border is smooth, its inferior border is sharp and thin with a **costal groove** on the inner surface against which lay intercostal nerves and blood vessels. Some variation exists in rib structure and articulations. The first rib is flattened superiorly-inferiorly and is quite broad. Ribs 1 and 10–12 only articulate with one vertebral body and ribs 11 and 12 do not articulate with the vertebral process, therefore lacking a tubercle. Fractures occur most commonly in the middle region, just anterior to the angle of the ribs. The first two ribs are protected by the clavicle, but fracture of the last two pair ("floating ribs") could cause damage to the kidneys, liver, or spleen.

Vertebral Column
(ver•**tee**•bral)

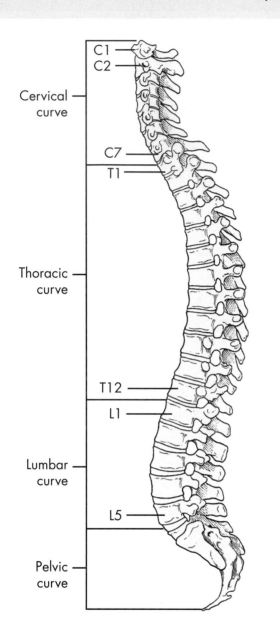

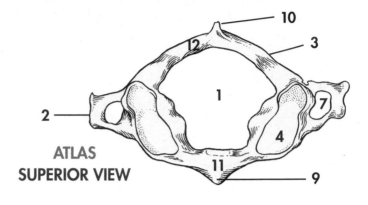

ATLAS
SUPERIOR VIEW

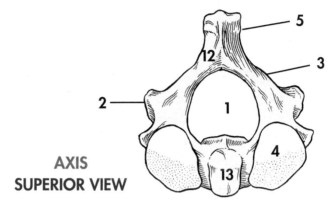

AXIS
SUPERIOR VIEW

TYPICAL CERVICAL
SUPERIOR VIEW

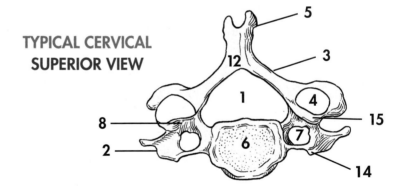

1 Vertebral (spinal) foramen	**5** Spinous process	**10** Posterior tubercle
	6 Body	**11** Anterior arch
2 Transverse process	**7** Transverse foramen	**12** Posterior arch
3 Lamina	**8** Pedicle	**13** Odontoid process (dens)
4 Superior articulating facet	**9** Anterior tubercle	

14 Anterior tubercle of transverse process	
15 Posterior tubercle of transverse process	

The **odontoid process** is actually the missing "body" of the atlas. The odontoid process acts as a pivot for the rotation of the atlas allowing side-to-side rotation of the head signaling "no." The anterior longitudinal ligament is severely stretched and may be torn during severe hyperextension of the neck causing a **whiplash** injury. There may also be a hyperflexion injury as the neck snaps back to the thorax. **Facet jumping** or locking of the cervical vertebrae may occur due to the dislocation of the facets.

Vertebral Column
(ver•**tee**•bral)

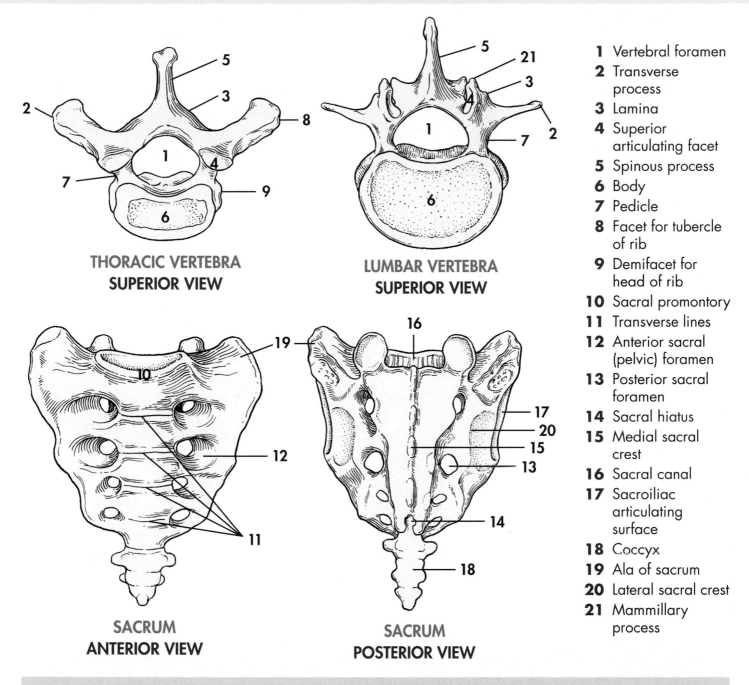

THORACIC VERTEBRA
SUPERIOR VIEW

LUMBAR VERTEBRA
SUPERIOR VIEW

SACRUM
ANTERIOR VIEW

SACRUM
POSTERIOR VIEW

1. Vertebral foramen
2. Transverse process
3. Lamina
4. Superior articulating facet
5. Spinous process
6. Body
7. Pedicle
8. Facet for tubercle of rib
9. Demifacet for head of rib
10. Sacral promontory
11. Transverse lines
12. Anterior sacral (pelvic) foramen
13. Posterior sacral foramen
14. Sacral hiatus
15. Medial sacral crest
16. Sacral canal
17. Sacroiliac articulating surface
18. Coccyx
19. Ala of sacrum
20. Lateral sacral crest
21. Mammillary process

Scheuermann's disease is a degeneration of adjacent vertebral bodies causing narrowing of intervertebral disk spaces and potential protrusion of intervertebral disks into the vertebral bodies, resulting in decreased spinal height and increased spinal curvature. High incidence occurs among gymnasts, trampolinists, cyclists, wrestlers, and rowers. Lower back and leg pain and motion restriction can be caused by a herniated or protruding intervertebral disk (**"slipped disk"**) and by enlargement of superior and inferior articulating facets due to repetitive stress and osteoarthritis (**facet syndrome**). Both conditions narrow the lateral foramina, pressing on the exiting spinal nerves. **Sciatica** is acute lower back pain that radiates down the posterolateral aspect of the thigh, often caused by a "slipped disk" at the L5–S1 level. The **sacral hiatus** is a common site for an epidural injection.

Vertebral Column Disorders and Injuries
(ver•**tee**•bral)

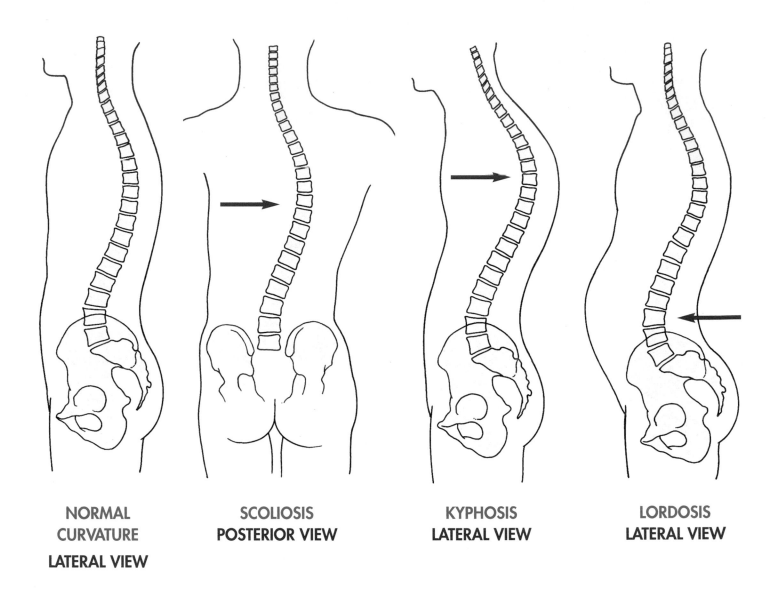

NORMAL CURVATURE	SCOLIOSIS	KYPHOSIS	LORDOSIS
LATERAL VIEW	**POSTERIOR VIEW**	**LATERAL VIEW**	**LATERAL VIEW**

There are four normal curvatures in the vertebral column. The cervical and lumbar curvatures are concave posteriorly, whereas the thoracic and sacral curvatures are convex posteriorly. These curvatures increase the flexibility of the vertebral column enabling it to function like a spring rather than a straight, rigid rod. In some people, the vertebral column may show **abnormal vertebral curvatures** caused by developmental or pathological processes. **Scoliosis** is a lateral curvature of one or more vertebral segments. More common in females, it becomes apparent in the teen years due to unequal growth of the two sides of the vertebral column caused by unequal development of vertebral muscles or of the sides of the vertebrae. **Kyphosis** is an exaggeration of the normal convex curvature of the thoracic vertebrae, producing a "humpback" or "Dowager's Hump" frequently related to osteoporosis. **Lordosis** is an exaggeration of the normal concave curvature of the lumbar region, frequently referred to as "swayback." During pregnancy, women often develop a temporary "lumbar lordosis" and lower back pain that is usually corrected at childbirth.

Right Clavicle
(klav•i•kal)

SUPERIOR VIEW

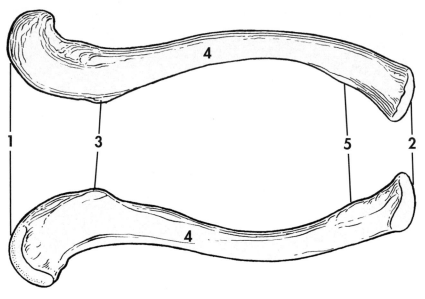

INFERIOR VIEW

1 Acromial extremity
2 Sternal extremity
3 Conoid tubercle
4 Body of clavicle
5 Costal tuberosity

Due to its S-shaped configuration, the **clavicle** is susceptible to compressive forces caused by a blow or fall on the point of the shoulder, by a direct blow to the clavicle, or by falling on an outstretched arm. Of concern in a fracture of the clavicle is the potential rupture of the subclavian vessels that lie just behind this bone. Fortunately, the natural anterior curvature of the clavicle in front of these vessels usually results in the clavicle fracturing forward rather than in a rearward direction. Typically, when the clavicle fractures, the **sternocleidomastoid muscle** elevates the medial bone fragment, and the arm drops because its weight prevents the **trapezius muscle** from holding up the lateral part of the fractured clavicle. In addition, the lateral fragment of the clavicle is pulled medially by the **adductors** of the arm, primarily the **latissimus dorsi** and **pectoralis major**. Sometimes, instead of causing a fracture, the same type of fall will cause an **AC joint sprain**. The **acromion process** is driven away from the **clavicle** resulting in a dislocation sometimes referred to as a separated shoulder.

Right Scapula
(**skap**•yoo•lah)

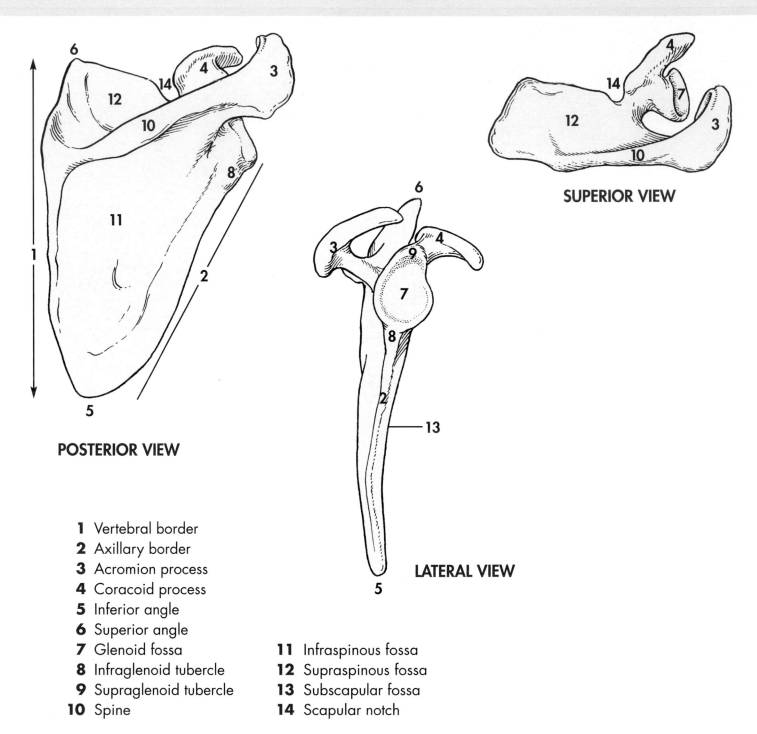

POSTERIOR VIEW

SUPERIOR VIEW

LATERAL VIEW

1 Vertebral border
2 Axillary border
3 Acromion process
4 Coracoid process
5 Inferior angle
6 Superior angle
7 Glenoid fossa
8 Infraglenoid tubercle
9 Supraglenoid tubercle
10 Spine

11 Infraspinous fossa
12 Supraspinous fossa
13 Subscapular fossa
14 Scapular notch

Compare the difference in depth between the **glenoid fossa** of the scapula and the **acetabulum** (pg. 18) of the os coxa. The shallower glenoid fossa permits a greater range of motion of the arm at the shoulder, but leaves the shoulder joint more likely to dislocate. This explains why the shoulder is more easily dislocated than the hip. **Scapular fractures** may occur in the main plate of the scapula, the **spine, acromion process, coracoid process,** or the **glenoid area. Avulsion fractures** to the coracoid process (page 30) result from direct trauma or forceful contraction of the **pectoralis minor** or short head of the **biceps brachii.**

Right Humerus
(hyoo•mir•us)

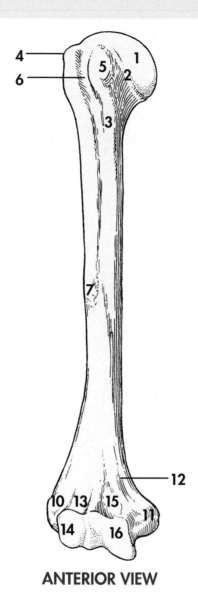

ANTERIOR VIEW

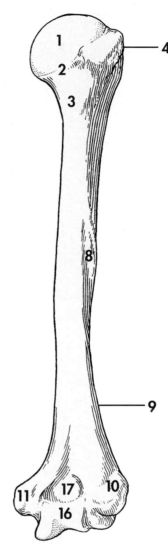

POSTERIOR VIEW

1 Head

2 Anatomical neck

3 Surgical neck

4 Greater tubercle

5 Lesser tubercle

6 Bicipital groove (intertubercular groove)

7 Deltoid tuberosity

8 Radial groove

9 Lateral supracondylar ridge

10 Lateral epicondyle

11 Medial epicondyle

12 Medial supracondylar ridge

13 Radial fossa

14 Capitulum

15 Coronoid fossa

16 Trochlea

17 Olecranon fossa

The **surgical neck** is the most common site for proximal humeral fractures. Also, because the **proximal humeral growth plate** does not close until the late teens, this is the site of fracture in youth commonly referred to as "**little league shoulder.**" Similarly, the partial or complete separation of the **medial epicondyle of the humerus** due to tension stress or repetitive, forceful contraction of the **flexor-pronator muscle group** causes "**little league elbow**" in youth. Fracture of the **secondary growth plate of the lateral epicondyle** is similar to the **medial epicondyle fracture.**

Right Radius and Ulna

(ray•de•us) (ul•nuh)

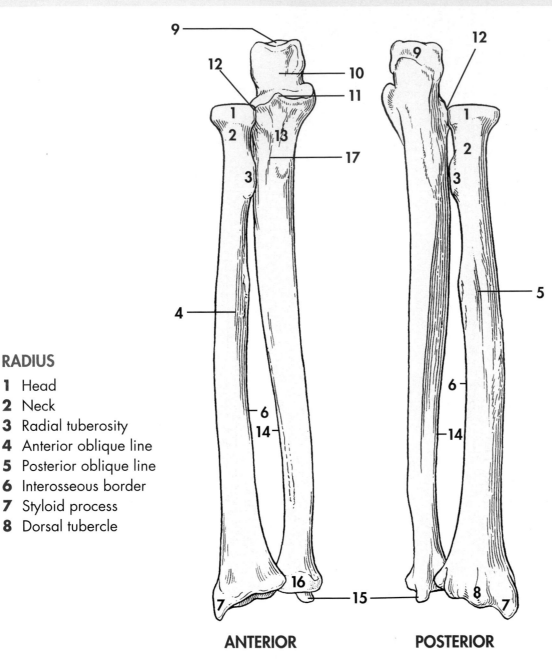

ANTERIOR **POSTERIOR**

RADIUS

1 Head
2 Neck
3 Radial tuberosity
4 Anterior oblique line
5 Posterior oblique line
6 Interosseous border
7 Styloid process
8 Dorsal tubercle

ULNA

9 Olecranon process
10 Trochlear (semilunar) notch
11 Coronoid process
12 Radial notch
13 Ulnar tuberosity
14 Interosseous border
15 Styloid process
16 Head
17 Supinator crest

In fractures of the distal radius or ulna, one or both bones may be fractured, or one bone fractured and the other bone dislocated at the wrist or elbow joint. In **Colles' fracture,** the fragment of a transverse fracture of the distal styloid process may impact with the shaft of the radius or be displaced dorsally and radially, producing a **"dinner-fork" deformity. Smith's fracture** is the reverse, with the fragment moving toward the palmar surface. In **Monteggia's fracture,** the ulna breaks and the radial head is dislocated; in **Galeazzi's fracture,** there is fracture of the radius and dislocation of the distal radioulnar joint.

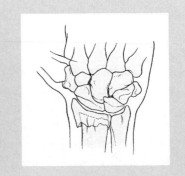

Right Hand

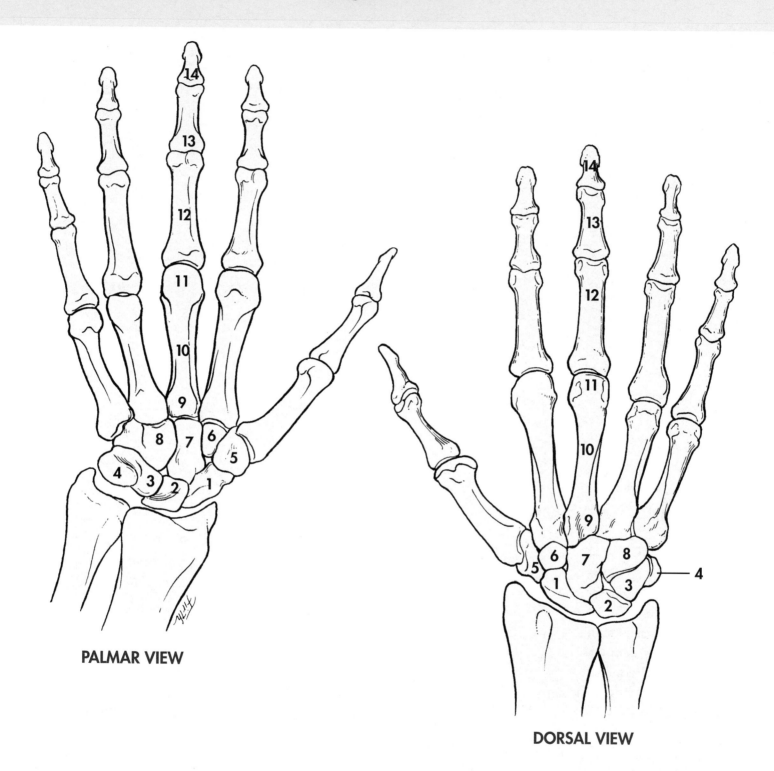

PALMAR VIEW

DORSAL VIEW

Carpals
Proximal Row
1 Scaphoid
2 Lunate
3 Triquetrum
4 Pisiform

Distal Row
5 Trapezium
6 Trapezoid
7 Capitate
8 Hamate

Metacarpals I through V
9 Base
10 Shaft
11 Head

Phalanges
12 Proximal
13 Middle
14 Distal

Right Os Coxa
(os **koks**•a)

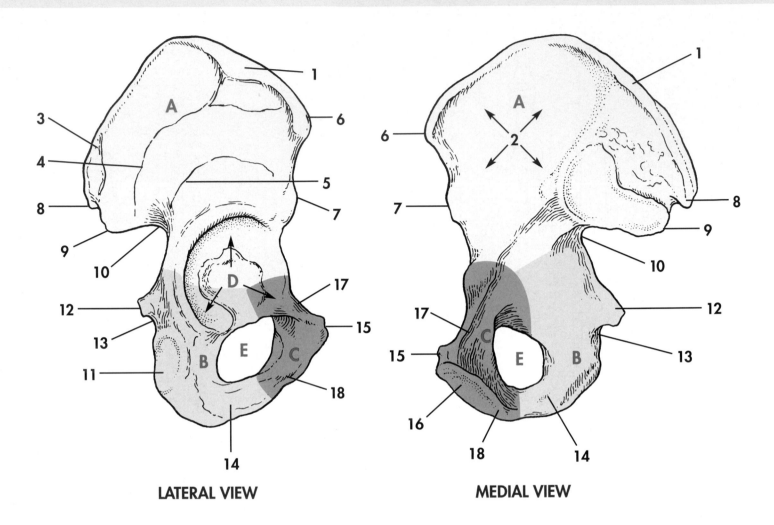

LATERAL VIEW

MEDIAL VIEW

A Ilium
1 Iliac crest
2 Iliac fossa
3 Posterior gluteal line
4 Anterior gluteal line
5 Inferior gluteal line
6 Anterior superior iliac spine
7 Anterior inferior iliac spine

8 Posterior superior iliac spine
9 Posterior inferior iliac spine
10 Greater sciatic notch

B Ischium
11 Ischial tuberosity
12 Ischial spine
13 Lesser sciatic notch
14 Ramus of ischium

C Pubis
15 Pubic crest
16 Pubic symphysis
17 Superior ramus of pubis
18 Inferior ramus of pubis

D Acetabulum

E Obturator Foramen

The adult skeleton typically consists of 206 bones. There are about 270 "bones" at birth with the number increasing and decreasing during development and adolescence. Most bones start out as multiple centers of ossification that later grow or "fuse" together to form a single adult bone. The time pattern in which bone centers appear and later grow together is responsible for the variation in the bone number. Most people are aware of this in the development of the "long bones" of their arms and legs, but did you know that the frontal bone and mandible are initially paired, that each os coxa is initially comprised of three separate bones, and that the sternum and sacrum are each made up of several fused bones?

Comparison of the Male and Female Pelvis
(pel•vis)

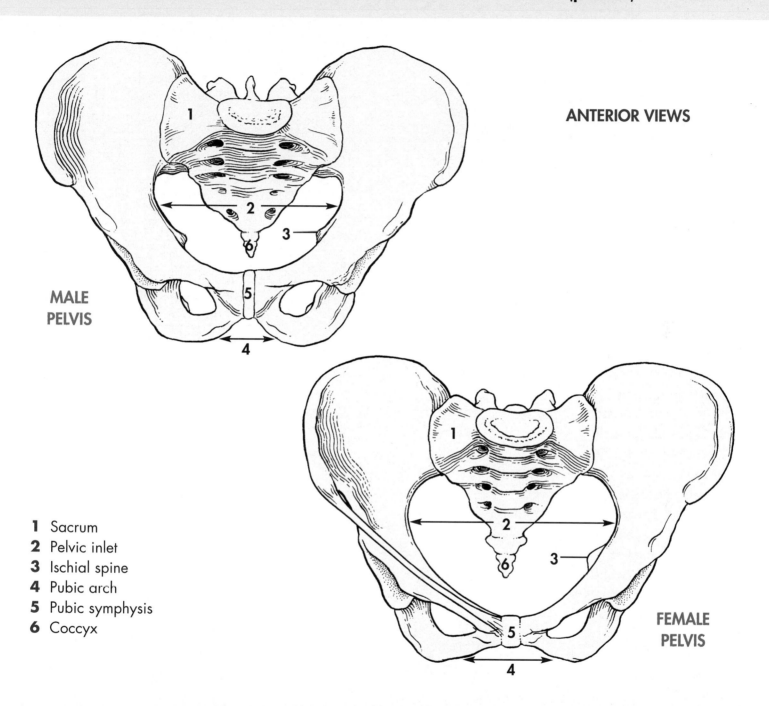

ANTERIOR VIEWS

MALE PELVIS

FEMALE PELVIS

1 Sacrum
2 Pelvic inlet
3 Ischial spine
4 Pubic arch
5 Pubic symphysis
6 Coccyx

There are several differences in the relative size and shape of parts of the female pelvis, which provide a wider passageway for childbirth. These consistent differences also enable one to identify the sex of a skeleton. The female pelvis is wider, shallower, and lighter. The hips (iliac crests) "flair" more in the female and the pelvic inlet is circular and proportionally larger than the heart-shaped inlet in the male. The pelvic outlet is increased in the female by a wider pubic arch (greater than 90 degrees) than in a male (less than 90 degrees). Also, in the female the **ischial tuberosities** are shorter and turned outward so the ischial spines point posteriorly. In the male, the ischial tuberosities are longer and point more medially as do the ischial spines. The wider and shorter female sacrum and coccyx curve in less sharply than do the longer and more curved male sacrum and coccyx.

Right Femur
(fee•mur)

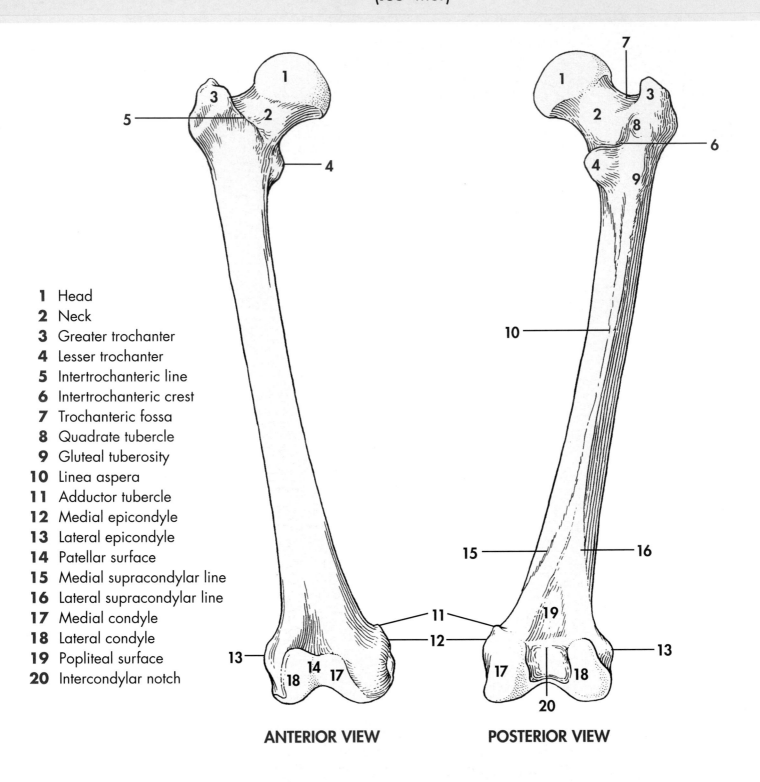

1 Head
2 Neck
3 Greater trochanter
4 Lesser trochanter
5 Intertrochanteric line
6 Intertrochanteric crest
7 Trochanteric fossa
8 Quadrate tubercle
9 Gluteal tuberosity
10 Linea aspera
11 Adductor tubercle
12 Medial epicondyle
13 Lateral epicondyle
14 Patellar surface
15 Medial supracondylar line
16 Lateral supracondylar line
17 Medial condyle
18 Lateral condyle
19 Popliteal surface
20 Intercondylar notch

ANTERIOR VIEW

POSTERIOR VIEW

Fractures of the femoral shaft can be very serious due to potential damage to adjacent nerves and blood vessels from bony fragments. Fractures are typically due to severe impact or compression forces. Fractures of the femoral neck are more common in the elderly and may require hip replacement surgery.

Right Tibia and Fibula
(**tib**•ee•ah)   (**fib**•yoo•lah)

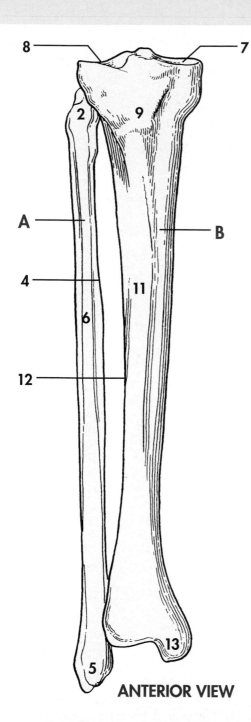

ANTERIOR VIEW

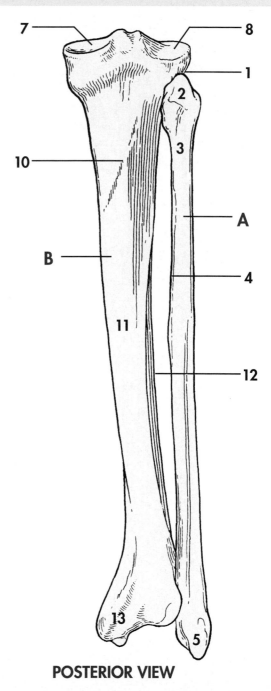

POSTERIOR VIEW

A Fibula
1 Apex of head
2 Head
3 Neck
4 Interosseous border
5 Lateral malleolus
6 Shaft

B Tibia
7 Medial condyle
8 Lateral condyle
9 Tibial tuberosity
10 Soleal line
11 Shaft
12 Interosseous border
13 Medial malleolus

The fibula is commonly fractured in skiing due to impact, and during soccer and basketball due to excessive inversion. A fracture 2 to 6 cm above the distal end of the lateral malleolus is called **Pott's fracture**. (See Chapter 3, page 51, for an illustration of Pott's fracture.) When the foot is extremely inverted, the ankle ligaments tear and the talus is forcibly tilted against the lateral malleolus, shearing it off. A **Maisonneuve fracture**, a fracture of the proximal third of the fibula, is typically due to extreme external rotation of the foot. Because the fibula is not crucial for walking, running, and jumping, it is a common source of a bone graft for reestablishing the blood supply and bone regeneration in a bone damaged due to trauma or removed because of malignancy.

Right Foot and Ankle

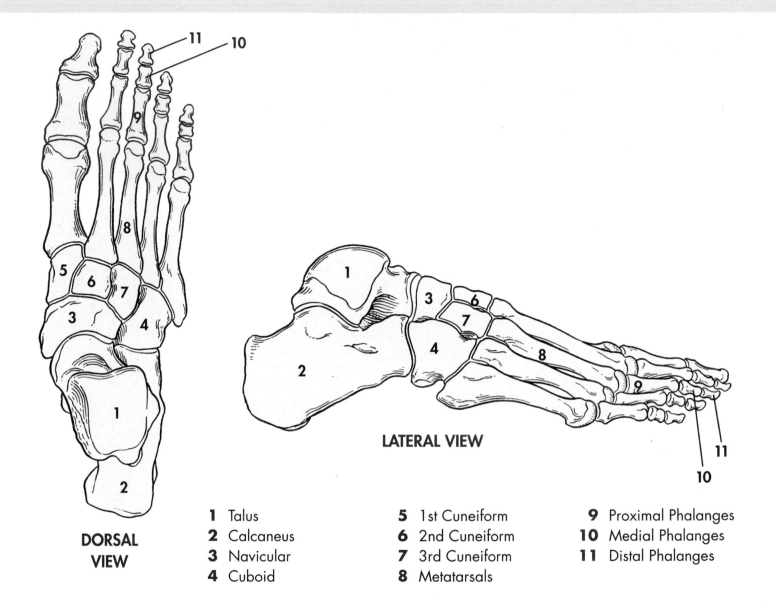

DORSAL VIEW

LATERAL VIEW

1	Talus	**5**	1st Cuneiform	**9**	Proximal Phalanges
2	Calcaneus	**6**	2nd Cuneiform	**10**	Medial Phalanges
3	Navicular	**7**	3rd Cuneiform	**11**	Distal Phalanges
4	Cuboid	**8**	Metatarsals		

LisFranc injuries include dislocations, ligament tears, and fractures associated with the bones of the "**middle foot**" forming the arch of the foot (**cuboid, cuneiforms and metatarsals**). Runners, skaters, and ballerinas are at risk of **stress fractures** in the metatarsals and navicular due to repetitive trauma. **Bone spurs**, a build up of bone, frequently occur at sites of stress. **Hammer toe** occurs when the proximal phalanx is permanently dorsiflexed at the metatarsal-phalanx joint, and the distal phalanx is plantar flexed, resulting in a hammer-like appearance of the toe. **Flat feet** and **fallen arches** occur because the plantar ligaments and fascia may be stretched beyond normal limits during standing, allowing the arch to fall. **Clubfeet** is a relatively common congenital defect in which the soles of the feet face medially and the toes point inferiorly. It can be corrected surgically. **Plantar fasciitis** is the most common hind foot problem in runners. Excessive tightness of the Achilles tendon or obesity can overload the **plantar fascia's** origin on the anteromedial aspect of the calcaneus. In a chronic condition, entrapment of the first branch of the lateral planter nerve contributes to the pain which is acute with the first steps in the morning. Pain is relieved with activity but recurs after rest.

Bony Landmarks and Palpations
Upper Body

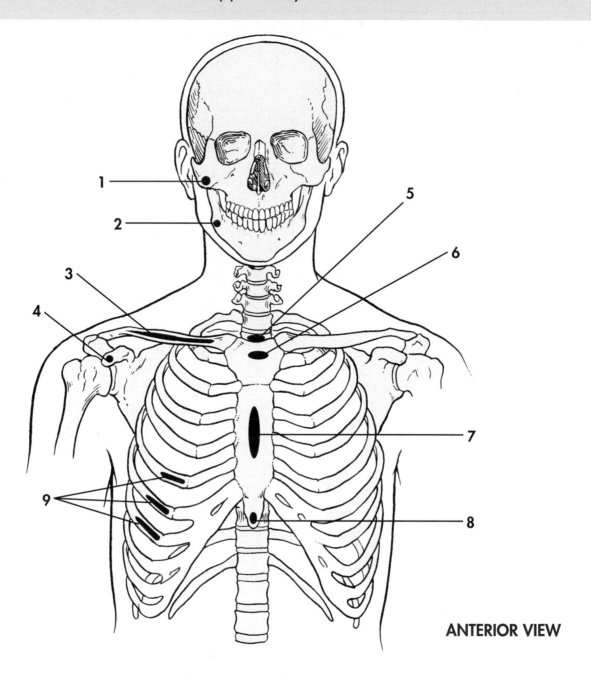

ANTERIOR VIEW

1 Palpate prominent **Zygomatic (cheek) bone** lateral to nose.

2 Palpate lower edge of jaw bone posteriorly to where it curves upward at the **Angle of jaw**.

3 Palpate the horizontal **Clavicle** from base of neck laterally toward shoulder joint.

4 At lateral end of clavicle, slide finger inferiorly and press to palpate the **Coracoid process of scapula**.

5 Palpate U-shaped **Jugular notch** at base of neck between medial ends of clavicles.

6 Palpate shield-shaped **Manubrium** immediately below jugular notch.

7 Palpate inferior to the manubrium along **Body of sternum**.

8 At inferior end of sternum, palpate the V-shaped **Xiphoid process**.

9 Palpate laterally from sternum along costal cartilages onto the body **Ribs**.

Bony Landmarks and Palpations
Upper Body—Posterior View

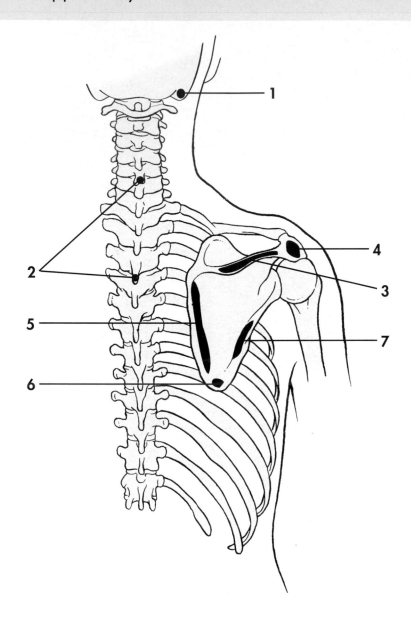

1 Press inward and upward behind the base of the ear lobe (pinna) to palpate the blunt **Mastoid process**.

On a partner in anatomical position:

2 With the person slightly bent forward, run fingers along the midline of the back from the top of the neck inferiorly to feel variations in height and terminal shape of the **Spinous processes of vertebrae**.

3 Palpate the ridge of bone, the **Spine of scapula**, from the medial to lateral edge of the scapula.

4 Palpate the prominent **Acromion process of scapula** at the lateral end of the scapula spine.

5 With arms medially rotated, palpate the **Vertebral border of scapula** lying parallel to the vertebral column.

6 Palpate inferiorly along the vertebral border of scapula to where it turns superior-laterally as the **Inferior angle of scapula**.

7 From the inferior angle, continue along the **Axillary border of scapula** toward the axillary (armpit) region.

Bony Landmarks and Palpations
Arm—Anterior and Posterior Views

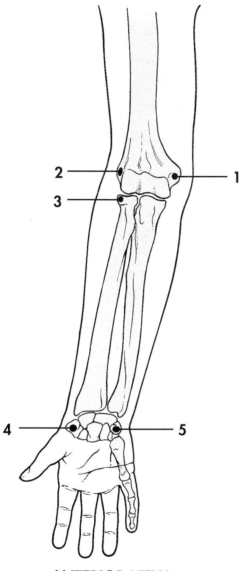

ANTERIOR VIEW

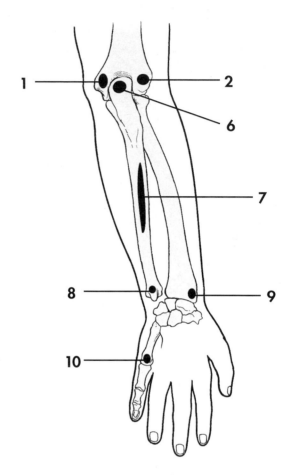

POSTERIOR VIEW

1,2 Place thumb and middle finger on medial and lateral "bumps" on distal end of humerus to palpate the **Medial and Lateral epicondyles of the humerus**.

3 In anatomical position, palpate the **Head of radius** on proximal end of radius just below lateral epicondyle of humerus.

4 Palpate **Scaphoid bone** on dorso-lateral side of wrist just distal to styloid process of radius. It forms the base of the "snuff-box."

5 Palpate **Pisiform bone** (small bony knob) just distal to styloid process of ulna on palmar wrist on the little finger side.

6 Palpate prominent **Olecranon process of ulna** on posterior surface of elbow.

7 Palpate sharp **Posterior border of ulna** along its length between olecranon and styloid processes.

8 Palpate prominent small, pointed **Styloid process of ulna** at distal end of ulna.

9 Palpate **Styloid process of radius** at distal end of radius.

10 Palpate **Heads of metacarpals ("knuckles")** by flexing metacarpal-proximal phalanges articulations.

Bony Landmarks and Palpations
Lower Body—Anterior View

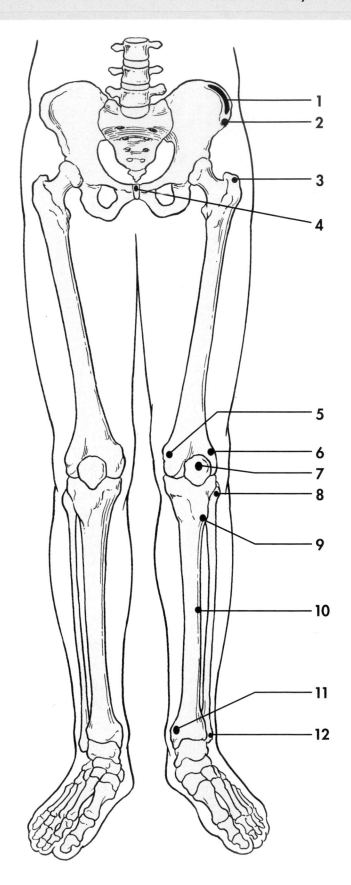

1 Palpate the bony **Iliac crest** at the superior surface of the ilium (hip bone).

2 Palpate along the iliac crest laterally and anteriorly to the **Anterior superior iliac spine** at its anterior end.

3 Palpate the large prominent **greater trochanter** at the superior lateral surface of the femur.

4 At the anterior midline of the pelvis, palpate the **Pubic symphysis**, the vertical ridge between the two pubic regions of the pelvis.

5,6 Place two fingers on each of the medial and lateral "bumps" on distal end of femur to palpate the **Medial and Lateral epicondyles of the femur**.

7 Palpate around the contour of the roughly triangular **Patella** (knee cap) overlapping the adjacent surfaces of the femur and tibia.

8 Palpate the **Head of fibula** at the proximal end of the fibula, lateral to and at same level as the tibial tuberosity.

9 Palpate the low, roughened **Tibial tuberosity** immediately below the inferior edge of the patella.

10 Palpate the sharp **Anterior edge of tibia** along its superior-inferior length.

11 Palpate the prominent **Medial malleolus of tibia** on the distal medial end of the tibia.

12 Palpate the prominent **Lateral malleolus of fibula** at the distal end of the fibula.

Bony Landmarks and Palpations
Lower Body—Posterior View

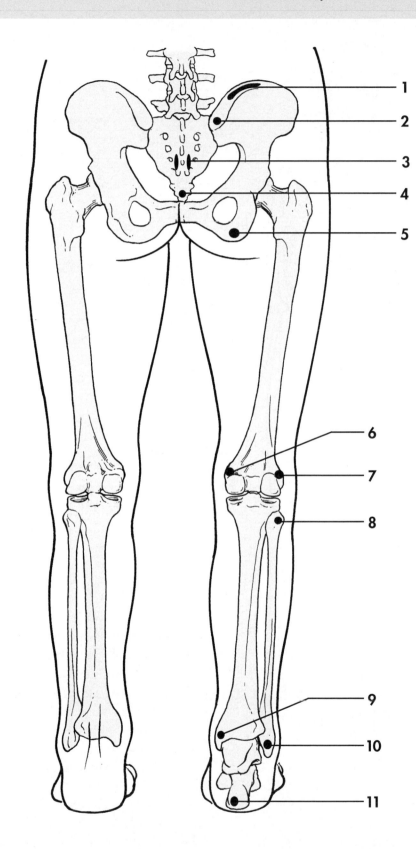

1 Iliac crest (pg. 26, #1)

2 Palpate along the iliac crest toward the sacrum and palpate the **Posterior superior iliac spine**.

3 Palpate between the posterior superior iliac spines to locate the **Medial sacral crest of the sacrum**.

4 Feel medially down to the end of the sacrum to palpate the small triangular bones of the **Coccyx**.

5 Palpate the large, blunt, and roughened process, the **Ischial tuberosity**, on which you sit.

6 **Medial epicondyle of femur** (pg. 26, #5)

7 **Lateral epicondyle of femur** (pg. 26, #6)

8 **Head of fibula** (pg. 26, #8)

9 **Medial malleolus of tibia** (pg. 26, #11)

10 **Lateral malleolus of fibula** (pg. 26, #12)

11 Palpate the large, posterior **Calcaneus** (heel bone) of the foot.

Fractures

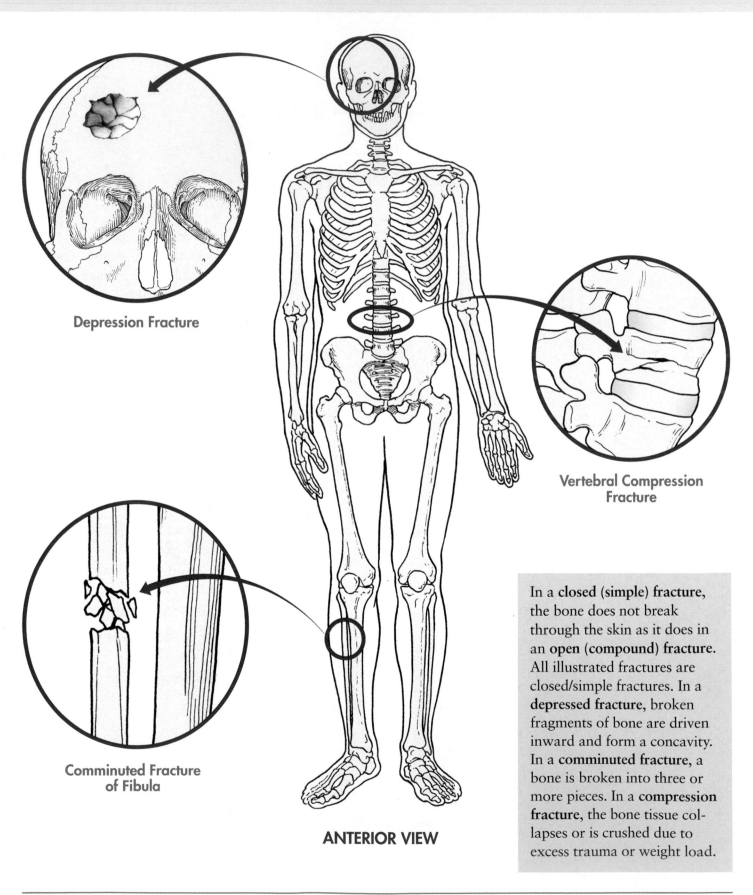

Depression Fracture

Vertebral Compression Fracture

Comminuted Fracture of Fibula

ANTERIOR VIEW

In a **closed (simple) fracture**, the bone does not break through the skin as it does in an **open (compound) fracture**. All illustrated fractures are closed/simple fractures. In a **depressed fracture**, broken fragments of bone are driven inward and form a concavity. In a **comminuted fracture**, a bone is broken into three or more pieces. In a **compression fracture**, the bone tissue collapses or is crushed due to excess trauma or weight load.

Fractures

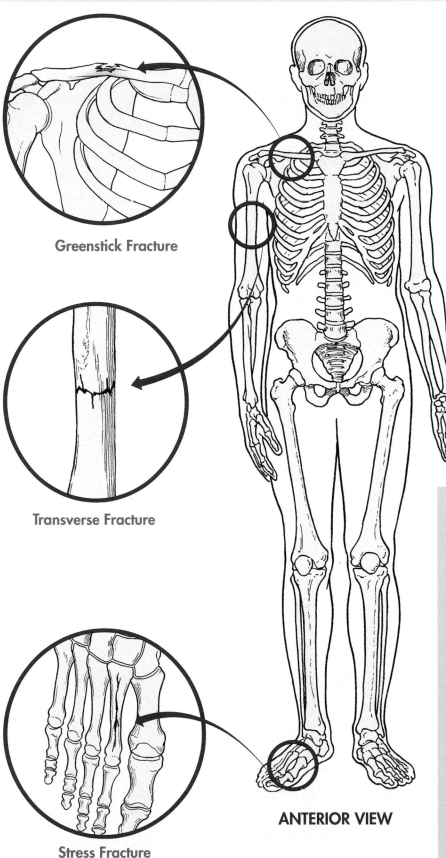

Greenstick Fracture

Transverse Fracture

Stress Fracture

ANTERIOR VIEW

In a **complete fracture**, the bone is broken into two or more separate pieces; in an **incomplete (partial/fissured) fracture** the break extends only partway through the bone as in the "greenstick" and "stress" fractures. A **greenstick fracture** is an incomplete break in which one side of the bone breaks and the other side bends. Greenstick fractures are more common in children than adults because developing bones have more organic than mineral matrix and are therefore more flexible. Multiple greenstick fractures are often a sign of child abuse. In a **transverse fracture**, a bone is broken horizontally across its width. In a **stress (fissured/partial) fracture**, there is a series of incomplete breaks parallel to the long axis of the bone.

Fractures

In a **nondisplaced fracture**, the fragments of the broken bone remain in anatomical alignment; in a **displaced fracture**, the bone fragments do not remain in correct anatomical alignment. All illustrated fractures are non-displaced fractures except for avulsion and spiral fractures. In an **avulsion fracture**, a portion of a bone is broken away, as a result of direct trauma or excessive muscle contraction against resistance. In an **oblique fracture**, the break occurs across the bone at an angle to the long axis of the bone. In an **impacted fracture**, one end of the broken bone is driven into the opposing portion of the bone. In a **spiral fracture**, due to excessive twisting, the fracture line "spirals around the bone." Because of the twisting of one part of the bone relative to the other, this type of fracture can be considered a **displaced fracture**.

Avulsion Fracture
(of coracoid process)

Impacted Fracture
(of femur)

Spiral Fracture of Tibia

POSTERIOR VIEW

Oblique Fracture
(of Tibia)

An Illustrated Atlas of the Skeletal Muscles

Youth Sports Epiphyseal Fractures
(ep•i•**fiz**•e•al)

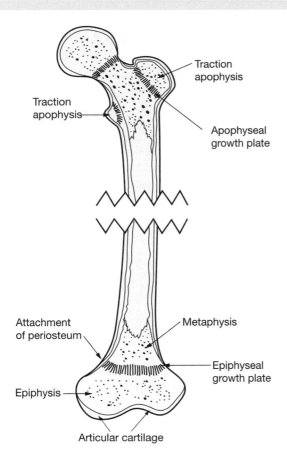

Traction apophysis

Traction apophysis

Apophyseal growth plate

Attachment of periosteum

Metaphysis

Epiphyseal growth plate

Epiphysis

Articular cartilage

Adolescence is the peak period for growth plate injuries in young athletes. Actively growing cartilage is found in the epiphyseal plates between the metaphysis and epiphysis of long bones, articular cartilages, and apophyseal plates deep to insertion sites of tendons. All of these cartilages may be subject to injury due to excessive and repetitive forces and are commonly referred to as "overuse injuries." Knowledge of the details of fractures is important for proper treatment of patients and assessing long-term effects.

Epiphyseal Fractures

Epiphyseal plates are responsible for growth in length; therefore, injury to epiphyseal plates may effect longitudinal growth of bones. Injuries at the ends of bones in the appendicular skeleton are commonly grouped according to the Salter-Harris Classification System (Types I–V) based on the pattern of damage to one or more of the epiphysis, epiphyseal growth plate and metaphysis.

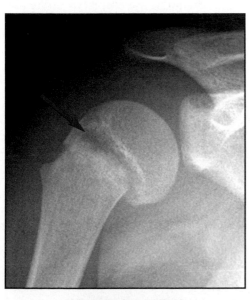

ANTERIOR VIEW

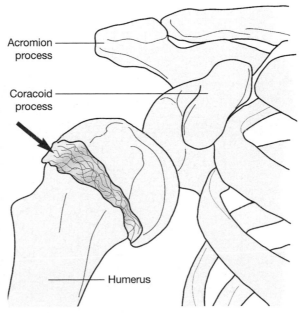

Acromion process

Coracoid process

Humerus

TYPE I is a partial or complete separation of the growth plate between the metaphysis and epiphysis (**arrows**) without any fracture of adjacent bone. Illustrated is the fracture of the proximal humeral epiphyseal plate referred to as "Little League Shoulder."

Youth Sports Epiphyseal Fractures
(ep•i•**fiz**•e•al)

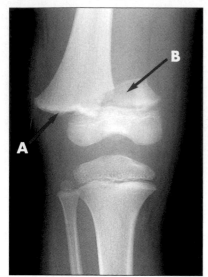

ANTERIOR VIEW

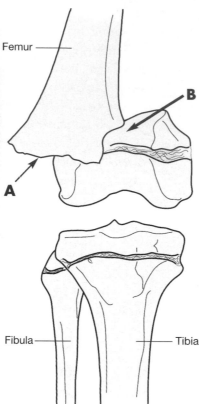

Femur

Fibula — Tibia

TYPE II is a partial separation of the growth plate (**arrow A**) with a bone fracture extending through the metaphysis (**arrow B**) on the side subjected to the excessive force. This is the most common fracture. The illustrated fracture of the distal end of the femur could result from a lateral force to the knee and lower leg in football or other contact sports.

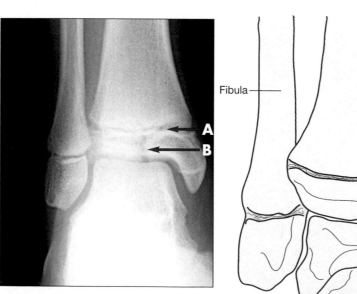

ANTERIOR VIEW

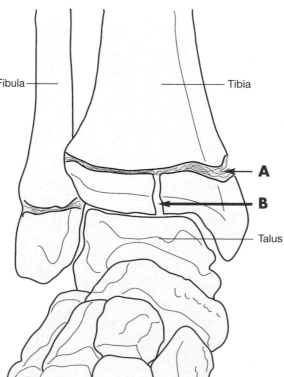

Fibula — Tibia

Talus

TYPE III is a partial separation of the growth plate (**arrow A**) with a bone fracture extending through the epiphysis (**arrow B**) into the joint, with possible damage to the articular cartilage. The illustrated fracture of the distal end of the tibia may occur as a result of twisting of the ankle in sports such as football and figure skating.

Youth Sports Epiphyseal Fractures
(ep•i•**fiz**•e•al)

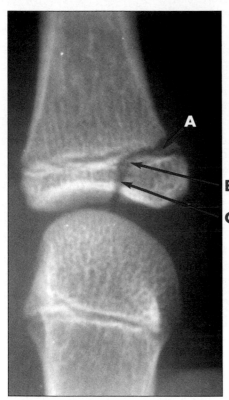

PALMAR VIEW

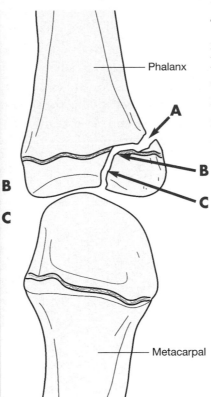

Phalanx

A

B

C

Metacarpal

TYPE IV is a fracture extending through the metaphysis (**arrow A**), growth plate (**arrow B**) and epiphysis (**arrow C**), with possible damage to the articular cartilage. Illustrated is a fracture of the finger that may occur due to jamming of the finger in sports such as volleyball and basketball.

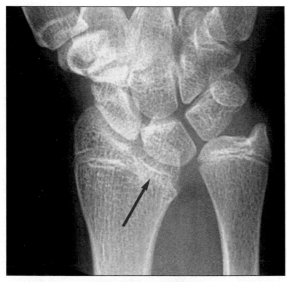

PALMAR VIEW

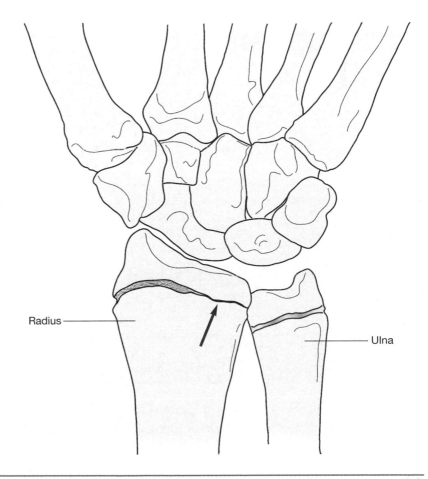

Radius

Ulna

TYPE V is a compression or crushing of part or all of the growth plate (**arrow**) due to excessive compression. It may effect longitudinal growth. Illustrated is the compression fracture of the distal radial epiphyseal plate, referred to as "gymnast's wrist."

Apophyseal Fractures

Apophyseal plate fractures are another group of overuse injuries. Apophyseal plates contribute to bone shape but not to longitudinal growth. As attachment sites for muscles to bones, apophyses are subject to strong muscle contraction. Due to an imbalance in the development of muscles and bones, excessive and repetitive contraction of muscles in active adolescents may partially or completely dislodge the superficial bony point of tendon attachment from the underlying growth plate. Osgood Schlatter and Sever's Disease are examples of apophyseal fracture injuries. Both types of injuries may occur in sports requiring sprinting and jumping.

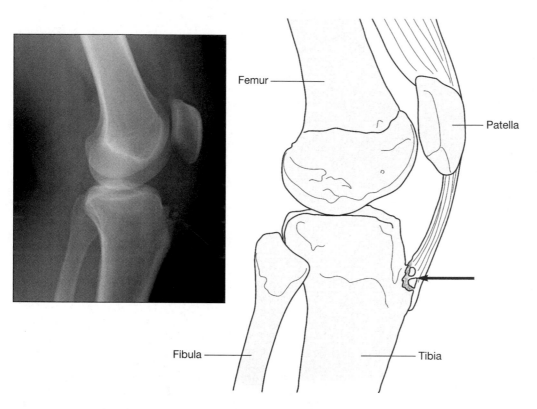

OSGOOD-SCHLATTER injury occurs when the tibial tuberosity is partially or completely separated from the underlying growth plate (**arrow**) due to excessive contraction of the quadriceps femoris muscle.

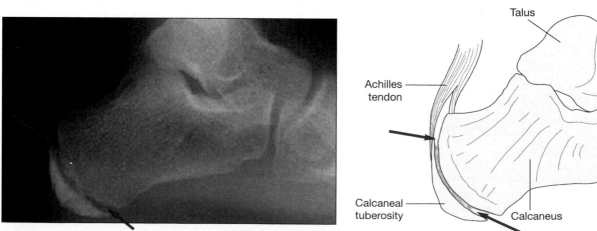

SEVER'S DISEASE, also known as "calcaneal apophysitis," occurs when the calcaneal tuberosity is partially or completely separated from the underlying growth plate (**arrows**) due to excessive contraction of the gastrocnemius muscle.

Articulations

2

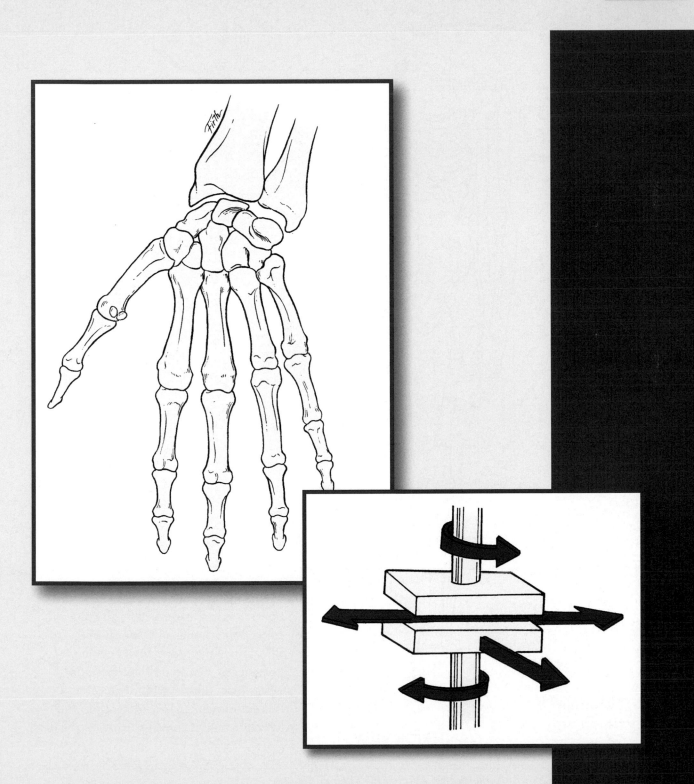

Fibrous Articulations
(**fy**•brus)(ar•**tik**•yoo•lay•shuns)

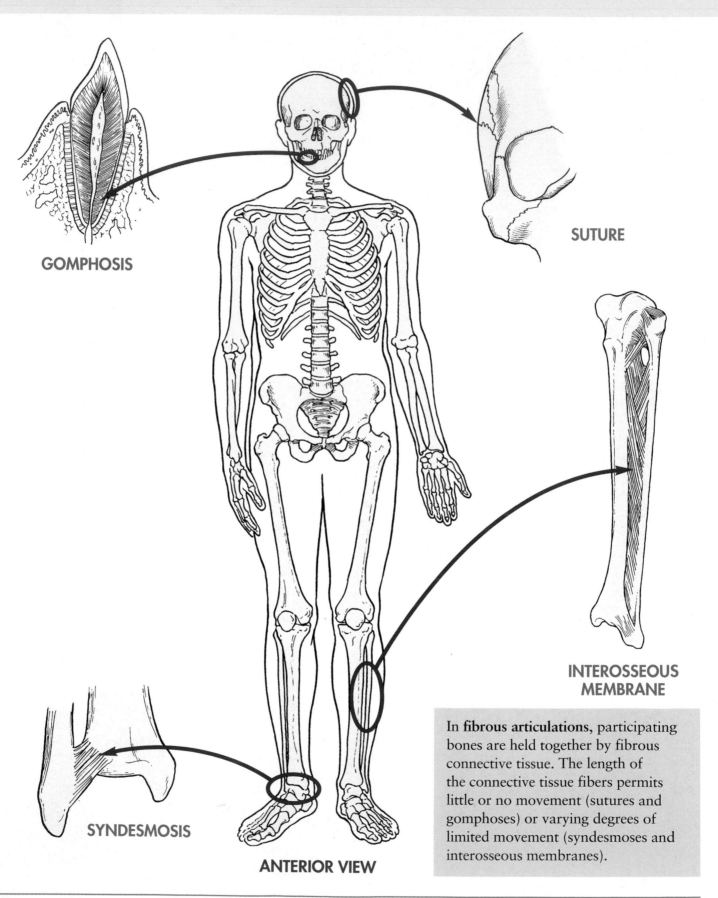

GOMPHOSIS

SUTURE

INTEROSSEOUS
MEMBRANE

SYNDESMOSIS

ANTERIOR VIEW

In **fibrous articulations**, participating bones are held together by fibrous connective tissue. The length of the connective tissue fibers permits little or no movement (sutures and gomphoses) or varying degrees of limited movement (syndesmoses and interosseous membranes).

An Illustrated Atlas of the Skeletal Muscles

Cartilaginous Articulations
(kar•ti•**laj**•in•nus)(ar•**tik**•yoo•lay•shuns)

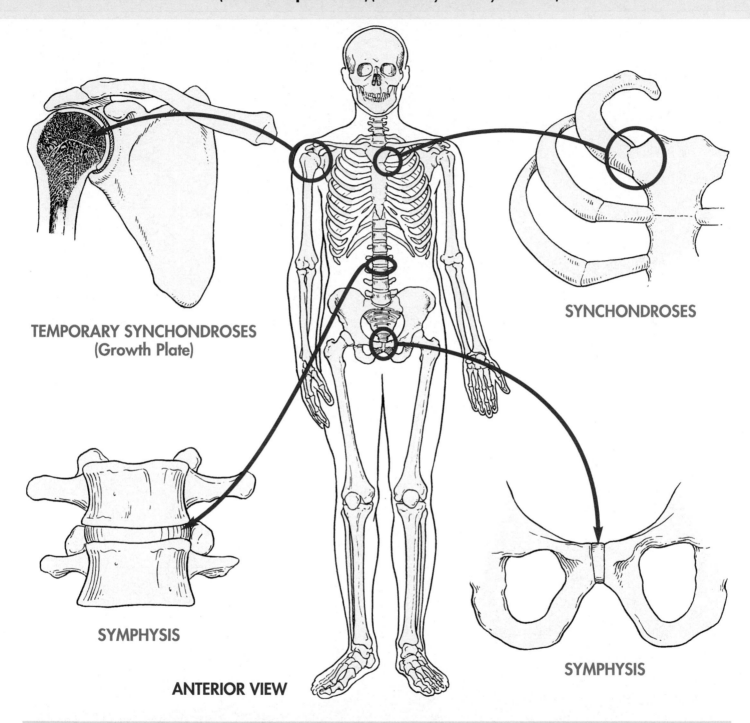

TEMPORARY SYNCHONDROSES
(Growth Plate)

SYNCHONDROSES

SYMPHYSIS

SYMPHYSIS

ANTERIOR VIEW

In these articulations bones are joined by either hyaline cartilage or fibrocartilage. In **synchondroses**, hyaline cartilage forms either temporary ("growth plates" in developing bones) or permanent (the first rib costal cartilage and the manubrium of the sternum) articulations.

Some texts still indicate that all rib/sternum joints are synchondroses, but close examination has determined that ribs 2–7 articulate with the sternum by gliding synovial (freely moveable) articulations. Fibro-cartilaginous **symphyses** are slightly moveable and provide strength and flexibility between the spool-shaped bodies of vertebrae ("intervertebral discs") and between the pubic portions of the pelvis (pubic symphysis).

Synovial Articulations
(si•**no**•ve•al)(ar•**tik**•yoo•lay•shuns)

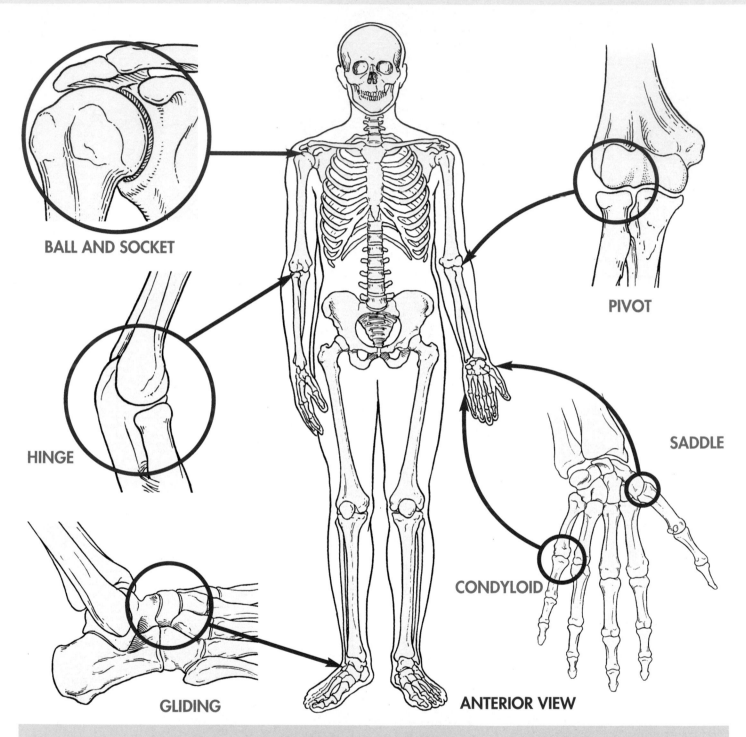

BALL AND SOCKET

PIVOT

HINGE

SADDLE

GLIDING

CONDYLOID

ANTERIOR VIEW

The majority of articulations are freely moveable **synovial joints.** The adjacent bony surfaces are capped by hyaline cartilage and joined by a fibrous articular capsule made up of an outer layer of ligaments and an inner synovial membrane that produces a viscous synovial fluid that fills the synovial cavity and lubricates the joint. Additional support is provided by surrounding ligaments and tendons. The direction and magnitude of movement allowed by each of the six subkinds of synovial joints is determined by the shapes of the apposing bone surfaces.

"Typical" Synovial Joint Structure

(si•**no**•ve•al)

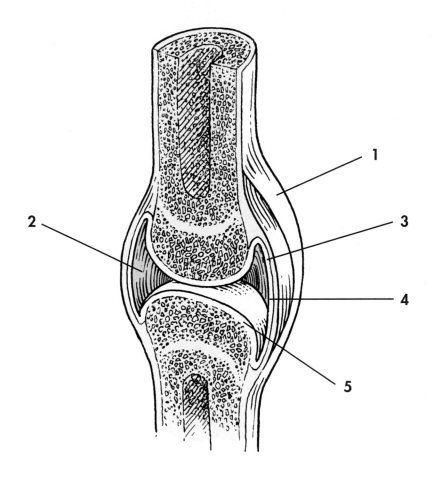

1 Ligament
2 Joint cavity
3 Fibrous capsule
4 Synovial membrane
5 Articular cartilage

The majority of articulations are freely moveable **synovial joints.** The adjacent bony surfaces are capped by hyaline cartilage and joined by a fibrous capsule made up of an outer layer of fibrous connective tissue and an inner synovial membrane that produces a viscous synovial fluid that fills the synovial cavity and lubricates the joint. Additional support is provided by surrounding ligaments and tendons. The direction(s) and amount of movement allowed by each of the six subkinds of synovial joints is determined by the shapes of the apposing bony surfaces.

Condyloid Subkind of Synovial Joints
(**kon**•de•loid)   (si•**no**•ve•al)

Metacarpal-Phalangeal Joint
(met•a•**kar**•pal) (fa•**lan**•gee•al)

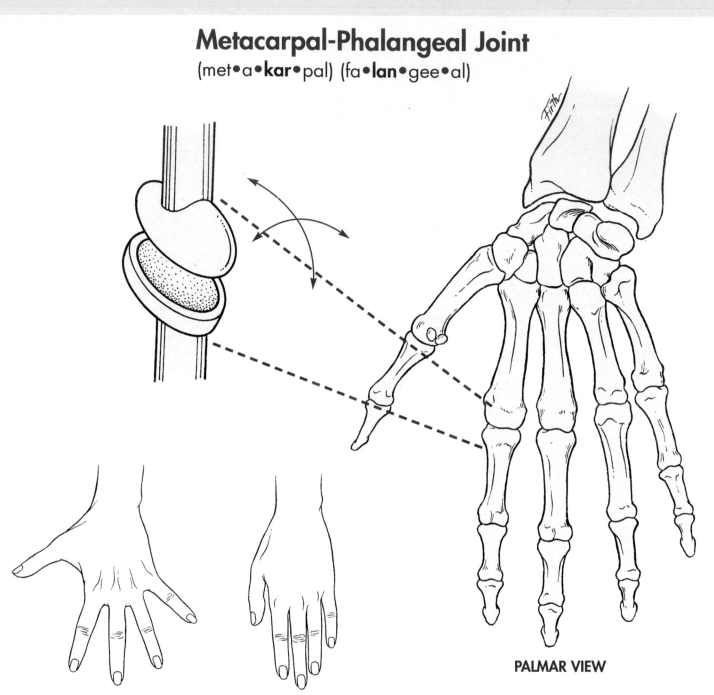

PALMAR VIEW

RANGE OF JOINT MOTION

Condyloid joints allow movement in two directions (**biaxial**). Simulate the movements of the hand diagrams by flexing or bending your fingers at the knuckles; then straighten and spread your fingers sideward. Other examples of condyloid joints are the occipital-atlas, temporal-mandibular (TMJ), radius-carpal, carpal-metacarpal, femur-tibia, tibia/fibula-talus, tarsal-metatarsal, and metatarsal-phalanges. Although many of these joints are often considered hinge joints because the most pronounced movement is flexion, they do exhibit movement in another direction as well.

Rotation Subkind of Synovial Joints
(si•**no**•ve•al)

Radius-Ulna Joint
(**ray**•de•us) (**ul**•nuh)

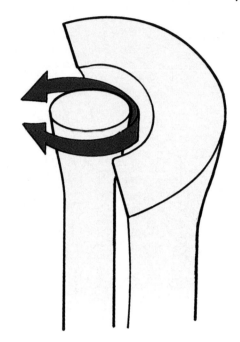

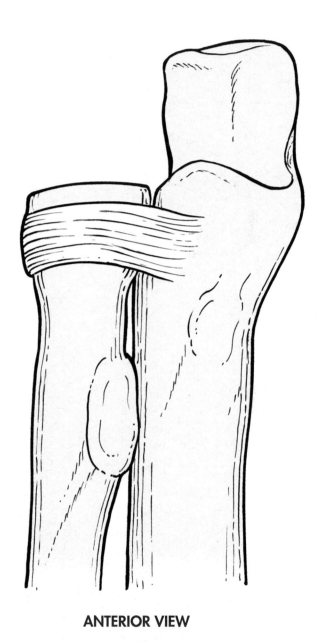

Atlas-Axis Joint

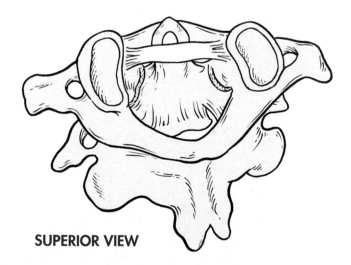

SUPERIOR VIEW

ANTERIOR VIEW

In a **pivot** or **rotational** joint, the only movement allowed is rotation around the longitudinal axis of the bone. That is why it is also called a **uniaxial joint**. Examples include the rotation of the **atlas** around the **odontoid process** of the **axis** and the proximal articulation between the **radius** and **ulna** as the radius rotates within the **annular ligament**.

Saddle Subkind of Synovial Joints
(si•**no**•ve•al)

Thumb Carpometacarpal Joint
(kar•po•met•a•**kar**•pal)

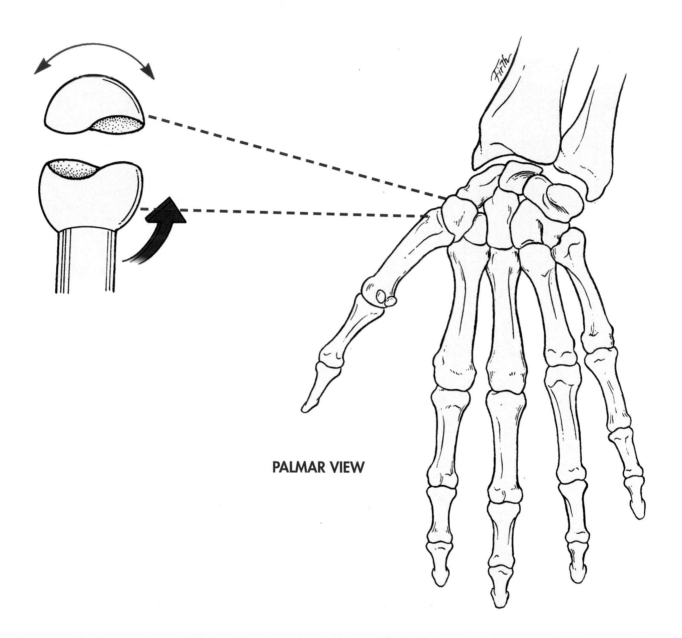

PALMAR VIEW

Saddle joints allow the same movements as condyloid joints—flexion, extension, abduction, adduction, and circumduction. The articular surface of each bone is concave in one direction and convex in the other. Therefore, the bones fit together as two English riding saddles would if they were rotated 90 degrees in relation to each other. The only true **saddle joint** is the **carpometacarpal joint** of the thumb. This flexible movement of the thumb enables you to grasp and hold objects.

Hinge Subkind of Synovial Joints
(si•**no**•ve•al)

Humerus-Ulna Joint
(**hyoo**•mir•us) (**ul**•nuh)

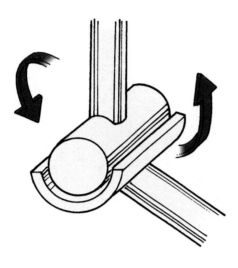

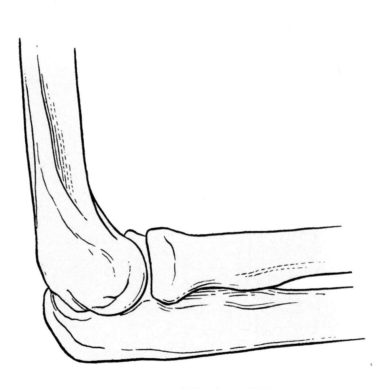

LATERAL VIEW

Hinge joints permit movement in only one direction, hence are referred to as **uniaxial**. The articular surfaces are shaped such that the only movements allowed are **flexion** and **extension**. The elbow, knee, and interphalangeal joints have long been considered examples of **hinge joints**. Today, there is controversy about the placement of the knee joint in this group.

Gliding Subkind of Synovial Joints
(si•**no**•ve•al)

Intercarpal and Intertarsal Joints
(in•ter•**kar**•pal)   (in•ter•**tar**•sal)

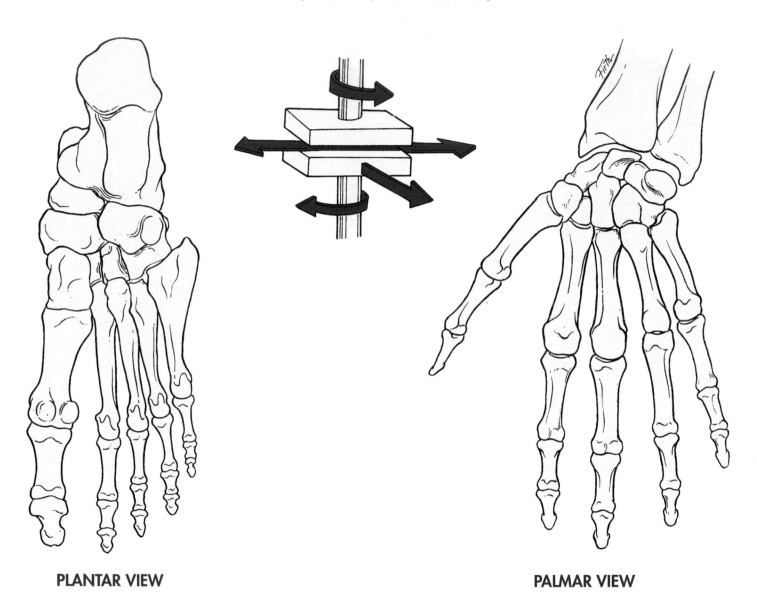

PLANTAR VIEW   **PALMAR VIEW**

Gliding joints are also called "plane" or "planar" joints. They are one of the most numerous types of joints in the body. The articular surfaces are flat or nearly so and best illustrated by the flat superior and inferior articulating surfaces or facets on the upper thoracic vertebrae. Even though the intercarpal and intertarsal surfaces are not as flat, they also illustrate gliding joints. These types of freely moveable joints allow short gliding movement in many directions, not around an axis, and thus are referred to as **nonaxial** joints. Other examples include vertebrocostal, sternocostal 2–7, and sterno- and acromioclavicular joints.

Ball and Socket Subkind of Synovial Joints
(si•**no**•ve•al)

Shoulder and Hip Joints

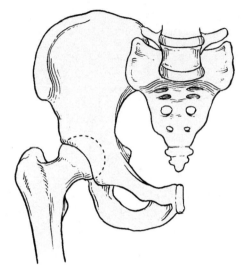

ANTERIOR VIEW

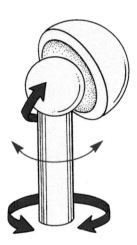

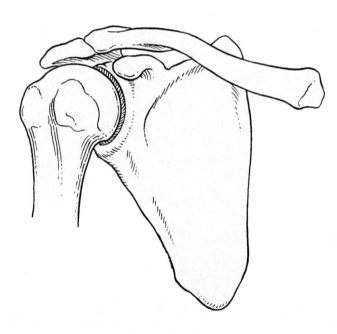

Ball and socket joints are formed by a spherical head of one bone fitting into a cup-shaped cavity on the other. Such joints allow movement around an indefinite number of axes and are also referred to as **multi-axial**. In addition to flexion, extension, abduction, adduction, and circumduction, ball and socket joints allow medial and lateral rotation. There are only two examples of ball and socket joints: the hip and the shoulder. The cup-shaped **acetabulum** on the hip is much deeper than the shallow **glenoid cavity** on the scapula, so the hip is more difficult to dislocate.

Shoulder Joint with Primary Ligaments

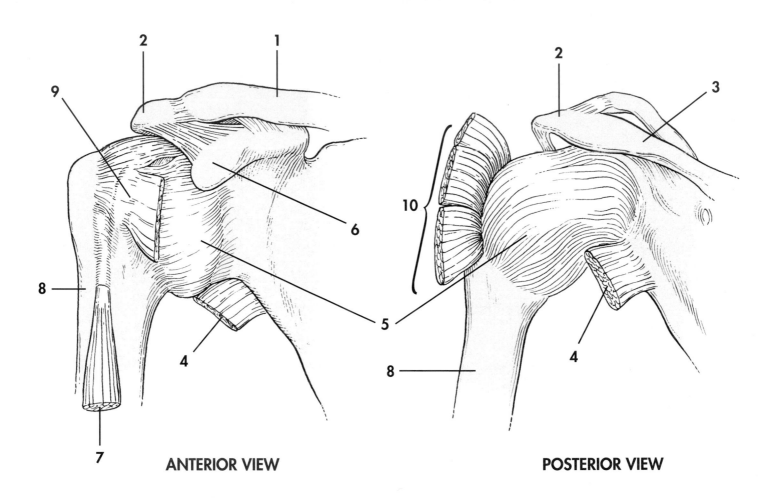

ANTERIOR VIEW

POSTERIOR VIEW

1 Clavicle
2 Acromion process
3 Spine of scapula
4 Triceps brachii (long head)
5 Articular capsule
6 Coracoid process
7 Biceps
8 Humerus
9 Subtendinous bursa
  of subscapular muscle
10 Supraspinatus

The shallow glenoid fossa for the head of the humerus, together with a loose and flexible joint capsule, provide for great freedom of movement at the shoulder, but not much stability. The muscles and tendons that cross the joint provide that strength and stability, particularly the tendon of the long head of the biceps brachii and the four muscles (supraspinatus, infraspinatus, subscapularis, and teres minor) that make up the **rotator cuff**. Extreme movements of the arm, such as by baseball pitchers, may tear or rupture one or more of the tendons of the rotator cuff muscles, especially the tendon of the supraspinatus muscle.

Elbow Joint with Ligaments

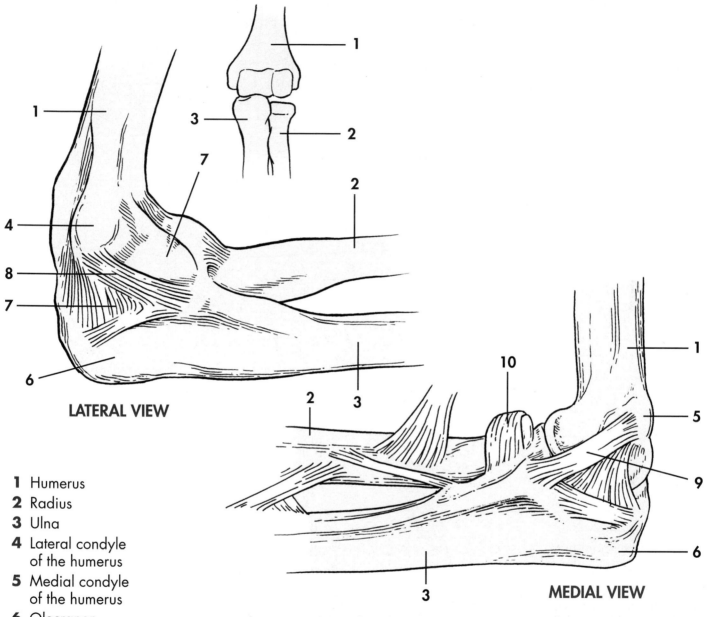

LATERAL VIEW

MEDIAL VIEW

1 Humerus
2 Radius
3 Ulna
4 Lateral condyle
   of the humerus
5 Medial condyle
   of the humerus
6 Olecranon
   process
7 Articular capsule
8 Radial collateral
   ligament
9 Ulnar collateral
   ligament
10 Annular ligament

The fibrous capsule completely encloses the joint. Its anterior and posterior parts are thin, but its sides are strengthened by the collateral ligaments. The fibrous capsule is attached to the proximal margins of the coronoid fossa anteriorly and the olecranon fossa posteriorly. Distally, the capsule is attached to the margins of the trochlear notch, the anterior border of the coronoid process, and the **annular ligament**. The **radial collateral ligament** is a strong band attached proximally to the lateral epicondyle of the humerus. The **ulnar collateral ligament** is triangular. It is composed of anterior and posterior bands and is attached to the **medial epicondyle** of the humerus. The strong anterior part is attached to the tubercle on the coronoid process of the ulna. The ulnar nerve is close to the ulnar collateral ligament.

Ligaments of Wrist and Hand

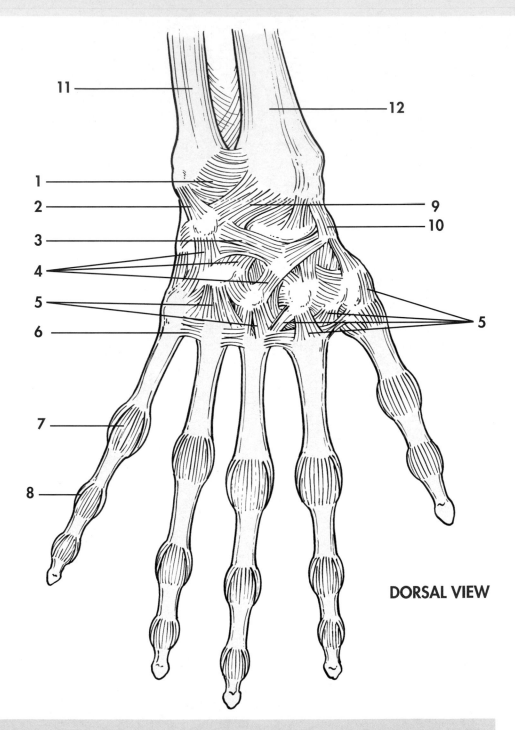

DORSAL VIEW

1 Dorsal radioulnar ligament
2 Ulnar collateral ligament
3 Dorsal carpal arcuate ligament
4 Dorsal intercarpal ligaments
5 Dorsal carpometacarpal ligaments
6 Dorsal metacarpal ligaments
7 Joint capsules
8 Collateral ligaments
9 Dorsal radiocarpal ligaments
10 Radial collateral ligament
11 Ulna
12 Radius

Numerous short ligaments provide stable but moveable connections between the bones of the wrist, the palm of the hand, and the phalanges. This ligamentous system is critical for facilitating the tremendous diversity of wrist and hand motions. The wrist joint is surrounded by a capsule strengthened by the palmar radiocarpal ligament, dorsal radiopalmar ligament, and the ulnar and radial collateral ligaments. The eight carpal bones are not tightly articulated with one another by their shapes, but are held together by interosseus ligaments and by palmar (volar), dorsal, radial, and ulnar ligaments. Other ligaments connect carpal bones to the proximal ends of the metacarpals.

Ligaments of Wrist and Hand

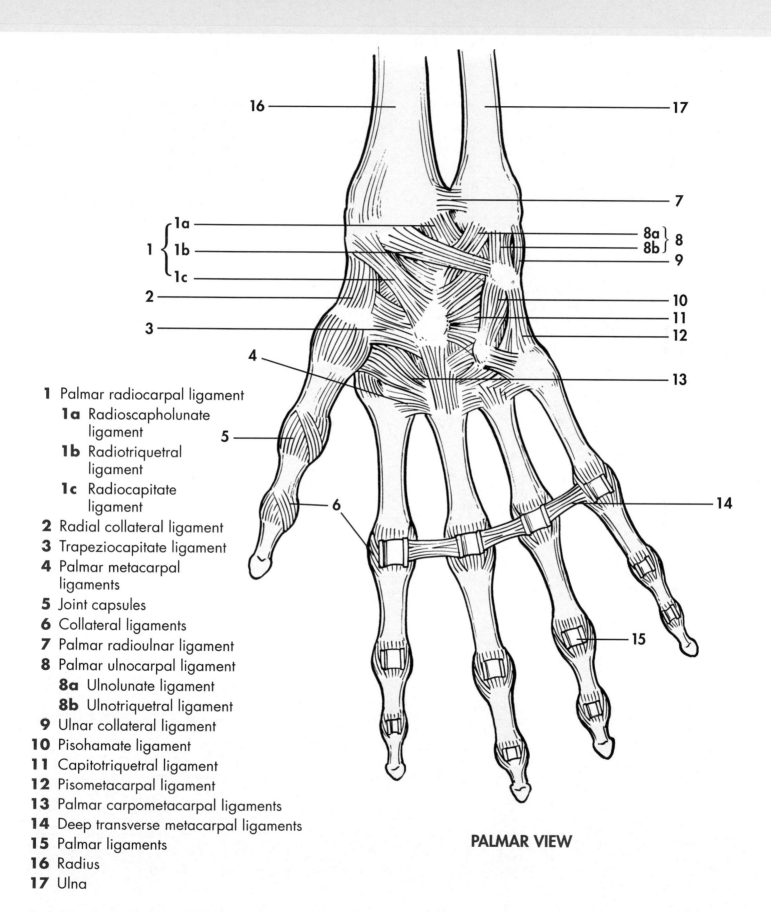

1 **Palmar radiocarpal ligament**
  1a Radioscapholunate ligament
  1b Radiotriquetral ligament
  1c Radiocapitate ligament
2 **Radial collateral ligament**
3 **Trapeziocapitate ligament**
4 **Palmar metacarpal ligaments**
5 **Joint capsules**
6 **Collateral ligaments**
7 **Palmar radioulnar ligament**
8 **Palmar ulnocarpal ligament**
  8a Ulnolunate ligament
  8b Ulnotriquetral ligament
9 **Ulnar collateral ligament**
10 **Pisohamate ligament**
11 **Capitotriquetral ligament**
12 **Pisometacarpal ligament**
13 **Palmar carpometacarpal ligaments**
14 **Deep transverse metacarpal ligaments**
15 **Palmar ligaments**
16 **Radius**
17 **Ulna**

PALMAR VIEW

Hip Joint with Primary Ligaments

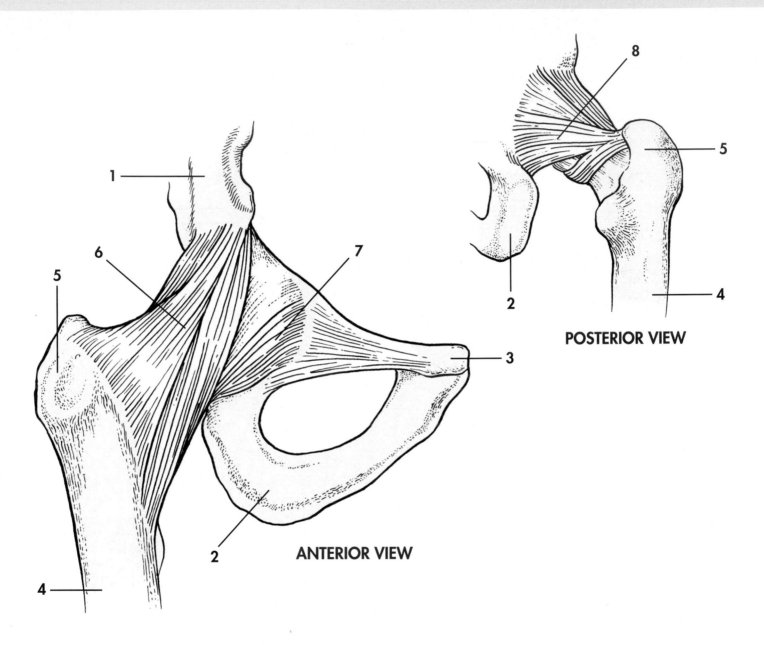

ANTERIOR VIEW

POSTERIOR VIEW

1 Ilium
2 Ischium
3 Pubis
4 Femur
5 Greater trochanter of femur
6 Iliofemoral ligament
7 Pubofemoral ligament
8 Ischiofemoral ligament

The **hip joint** is a ball and socket joint. The head of the **femur** fits into the acetabulum of the os coxa. The ligamentum teres attaches to a fovea or pit in the head of the femur. The **iliofemoral ligament** is a strong band that covers the anterior aspect of the hip joint. It attaches to the anterior inferior iliac spine and the acetabular rim. The **pubofemoral ligament** arises from the pubic part of the acetabular rim and blends with the medial part of the iliofemoral ligament. It strengthens the inferior and anterior parts of the joint. The **ischiofemoral ligament** arises from the ischial portion of the acetabular rim and spirals to the neck of the femur, medial to the base of the greater trochanter. It prevents hyperextension of the hip joint.

Knee Joint with Primary Ligaments

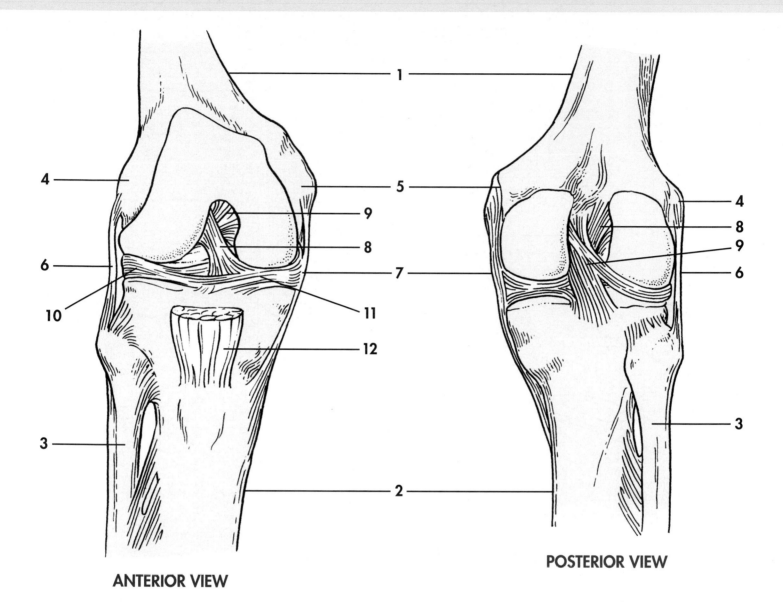

ANTERIOR VIEW

POSTERIOR VIEW

1 Femur
2 Tibia
3 Fibula
4 Lateral condyle of femur
5 Medial condyle of femur
6 Lateral (fibular) collateral ligament
7 Medial (tibial) collateral ligament
8 Anterior cruciate ligament
9 Posterior cruciate ligament
10 Lateral meniscus
11 Medial meniscus
12 Patella (tendon) ligament

The knee is the most complicated joint in the body. It is not as stable as other joints, and it is one of the most often used and damaged. Standing in one position with the knees slightly flexed can cause damage to the articular cartilage. The **menisci** act as shock absorbers. If the movement against them is too abrupt, they can be crushed or torn. The two **collateral ligaments** give stability to the joint. The **anterior** and **posterior cruciate ligaments** give additional front-to-back stability. Tearing of the anterior cruciate ligament (ACL) is a common sports injury.

Ligaments of Ankle and Foot

MEDIAL VIEW

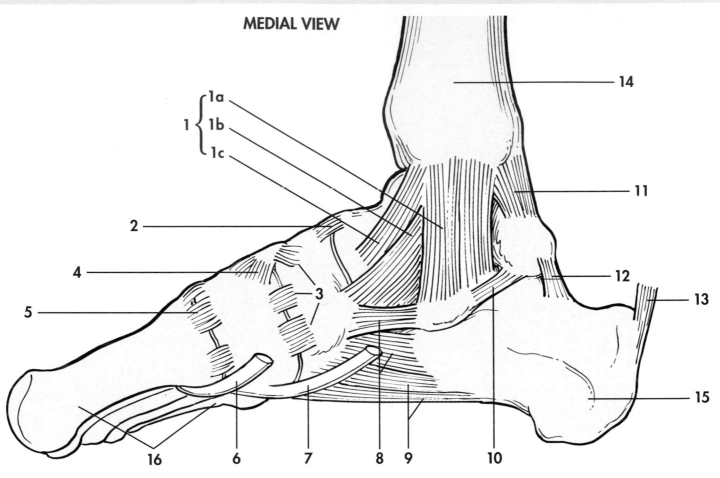

1 Deltoid ligaments	**8** Plantar calcaneonavicular ligaments
1a Tibiocalcaneal ligament	**9** Long plantar ligament
1b Tibionavicular ligament	**10** Medial talocalcaneal ligament
1c Anterior Tibiotarsal ligament	**11** Posterior tibiotalar ligament
2 Dorsal talonavicular ligament	**12** Posterior talocalcaneal ligament
3 Dorsal cuneonavicular ligaments	**13** Achilles tendon
4 Intercuneiform ligament	**14** Tibia
5 Dorsal tarsometatarsal ligament	**15** Calcaneus
6 Tibialis anterior tendon	**16** Metatarsals
7 Tibialis posterior tendon	

The ankle/foot is a complex joint with many ligaments that hold tendons in place, stabilize and hold bones together, and facilitate a variety of motions while also preventing excessive movements in other directions. The talus is normally held in place by the adjacent malleoli ("stabilizing") processes of the fibula and tibia. When the distance between the talus and the malleoli becomes larger, such as when the foot is extended forward, the stability of the ankle joint is more dependent on the ligaments extending from the malleolar processes to various ankle and foot bones. Laterally, four ligaments (anterior tibiofibular, anterior talofibular, posterior talofibular, calcaneofibular) extend from the fibula. Medially, four ligaments (tibiocalcaneal, tibionavicular, anterior tibiotalar, posterior tibiotalar) extend from the tibia.

Ligaments of Ankle and Foot

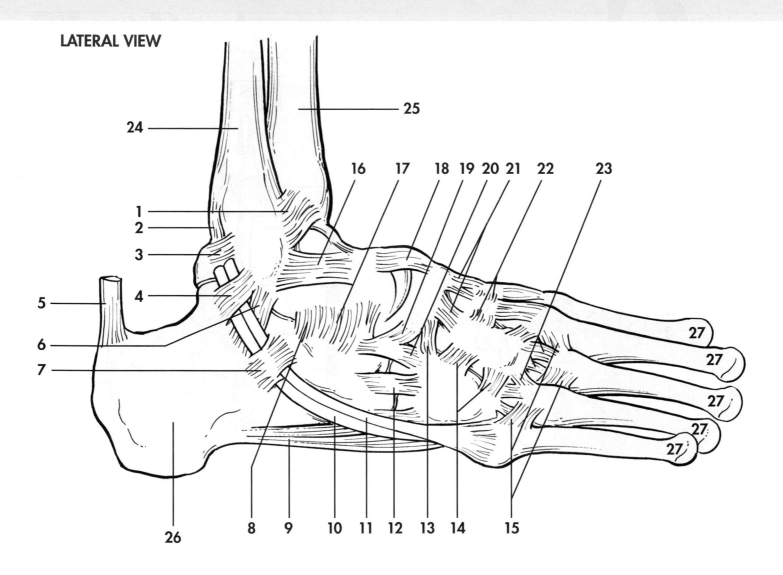

LATERAL VIEW

1 Anterior tibiofibular ligament
2 Posterior tibiofibular ligament
3 Posterior talofibular ligament
4 Superior peroneal retinaculum
5 Achilles tendon
6 Calcaneofibular ligament
7 Inferior peroneal retinaculum
8 Lateral talocalcaneal ligament
9 Longitudinal plantar ligament
10 Fibularis longus tendon
11 Fibularis brevis tendon
12 Dorsal calcaneocuboid ligament
13 Dorsal cuboideonavicular ligament
14 Dorsal cuneocuboid ligaments

15 Dorsal metatarsal ligaments
16 Anterior talofibular ligament
17 Interosseous talocalcaneal ligament
18 Dorsal talonavicular ligament
19 Calcaneonavicular ligament
20 Calcaneocuboidal ligament
21 Dorsal cuneonavicular ligaments
22 Dorsal intercuneiform ligaments
23 Dorsal tarsometatarsal ligaments
24 Fibula
25 Tibia
26 Calcaneus
27 Metatarsals

Flexion—Extension—Hyperextension
(flek•shun) (ex•sten•shun) (hi•per•ex•sten•shun)

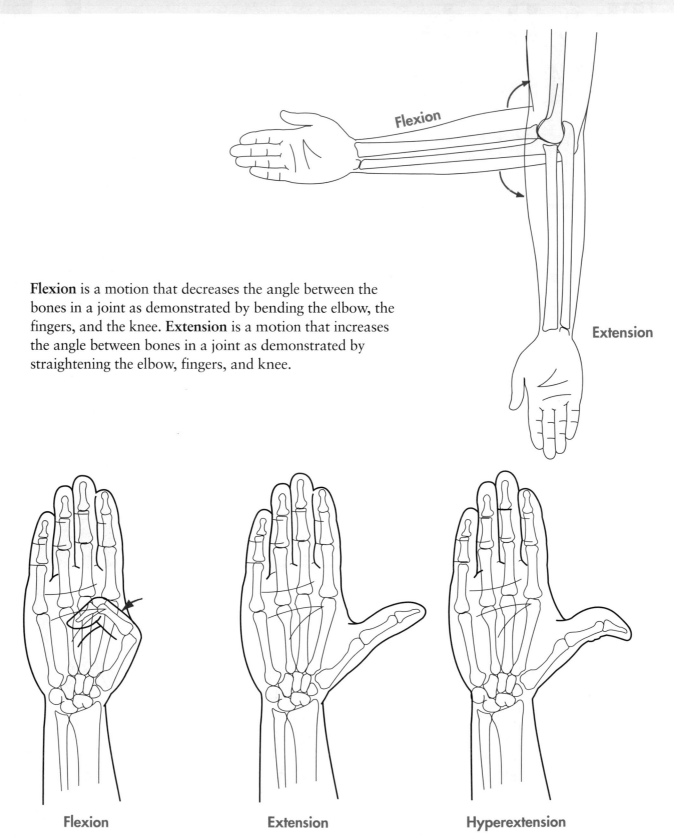

Flexion is a motion that decreases the angle between the bones in a joint as demonstrated by bending the elbow, the fingers, and the knee. **Extension** is a motion that increases the angle between bones in a joint as demonstrated by straightening the elbow, fingers, and knee.

Flexion

Extension

Hyperextension

Hyperextension is a motion that goes beyond its normal limits, as seen in the "hitchhiker's thumb." Bending the head backward is also an example of hyperextension.

Abduction and Adduction
(**ab**•duck•shun) (**ad**•duck•shun)

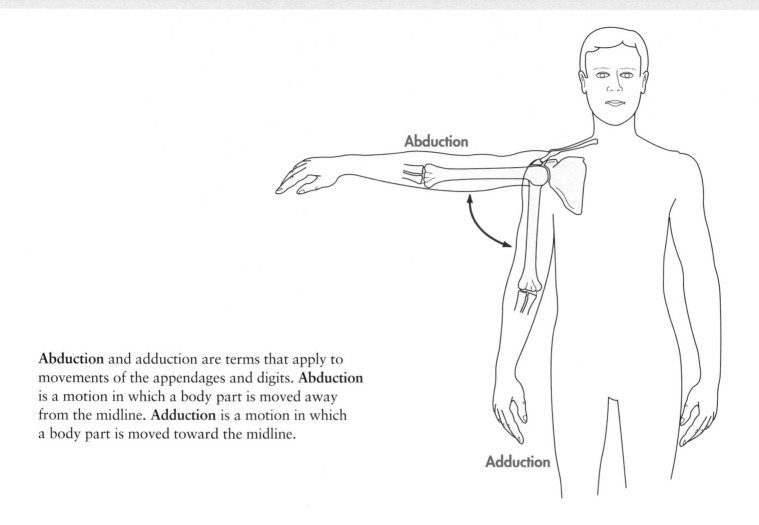

Abduction and adduction are terms that apply to movements of the appendages and digits. **Abduction** is a motion in which a body part is moved away from the midline. **Adduction** is a motion in which a body part is moved toward the midline.

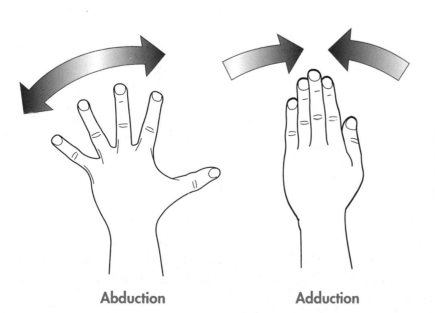

In reference to the hands and feet, abduction and adduction are defined as movement away from or toward the middle digit.

Abduction   **Adduction**

Rotation

(ro•**tay**•shun)

Atlas–Axis Articulation

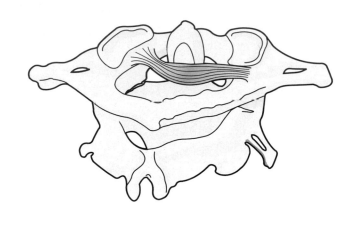

Os Coxa–Femur Lateral/Medial Rotation

(os **koks**•a) (**fee**•mur)

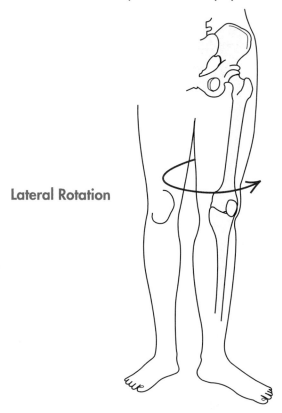
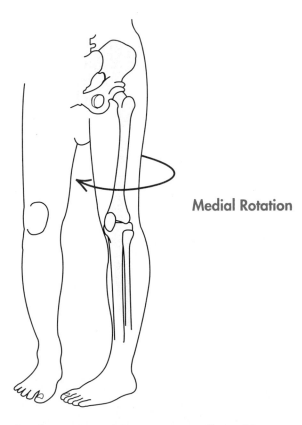

Lateral Rotation

Medial Rotation

Rotation is a motion in which a bone turns around its own longitudinal axis. It is the movement allowed between the first two cervical vertebrae, involving rotation of the atlas around the odontoid process (dens) of the axis, producing the characteristic side-to-side "no" motion of the head. Medial and lateral rotation of the arm and leg occur at the shoulder and hip joints.

Pronation and Supination
(pro•**nay**•shun)   (soo•pin•**nay**•shun)

Radius–Ulna Articulations
(**ray**•de•us) (**ul**•nuh)

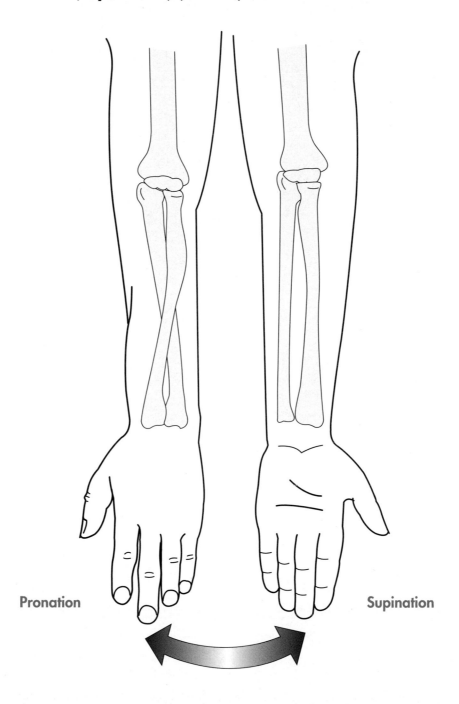

Pronation

Supination

Pronation and **Supination** are special rotational movements of the forearm that respectively cause the palm of the hand to face posterior or anterior (anatomical position). These motions are permitted by rotation between the radius and ulna at both the proximal and distal ends of the two bones. In anatomical position (supination), the two lower arm bones are parallel; in pronation, the radius lies diagonally across the ulna.

Circumduction
(sur•kum•**duck**•shun)

Scapular–Humerus (Shoulder) Joint
(**skap**•yoo•lahr)(**hyoo**•mir•us)

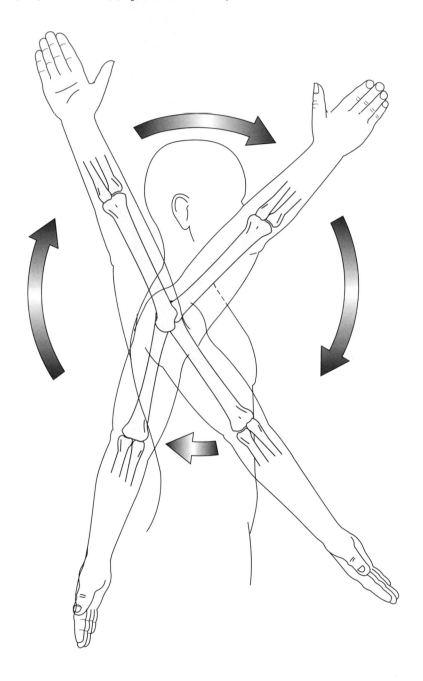

Circumduction is a motion that occurs at the "ball and socket" joints at the shoulder and the hip. In circumduction, the movement of the limbs describes a cone in space. The distal end of the limb moves in a circle while the proximal end is more or less stationary. The 360° rotation of circumduction includes flexion, abduction, extension, and rotation and is the quickest way to exercise the many muscles that move the hip and shoulder. A pitcher winding up to throw a baseball is circumducting the pitching arm.

Plantar Flexion—Dorsiflexion
(**plan**•tahr)(**flek**•shun) (door•se•**flek**•shun)

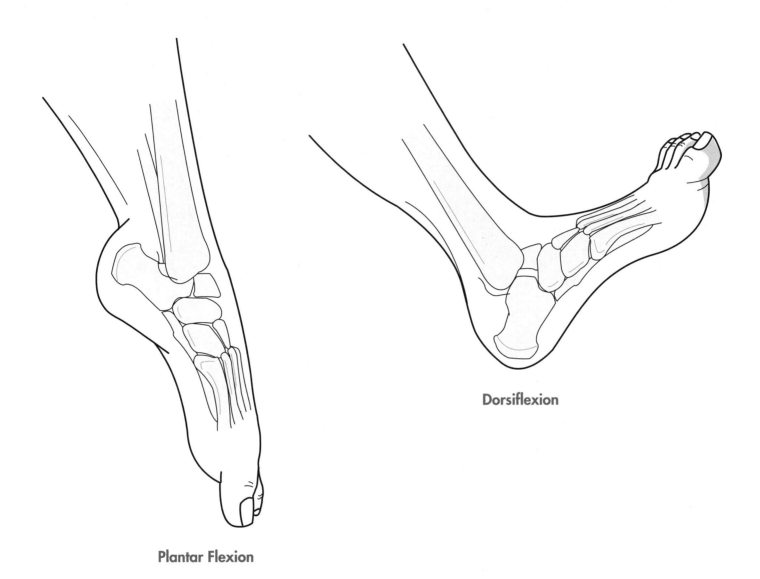

Dorsiflexion

Plantar Flexion

Plantar flexion is a downward movement of the foot and toes at the ankle resulting in the foot and toes pointing toward the floor, such as in the "toe pointed" position used by gymnasts. In older textbooks and lab manuals, plantar flexion was referred to as plantar extension. **Dorsiflexion** is an upward movement of the foot and toes at the ankle, resulting in the toes and foot projecting away from the floor.

Opposition—Reposition
(op•poh•zih•shun)(re•poh•zih•shun)

Thumb–Finger Touching

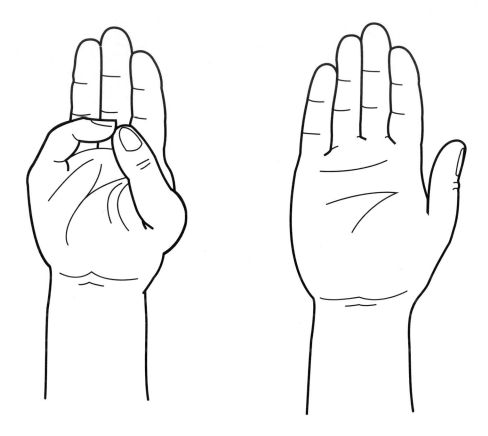

Opposition is movement of the thumb to approach or touch one or more of the fingertips; **reposition** is the reverse movement, returning the thumb to a parallel position with the fingers.

Since opposition is the movement that enables the hand to grasp objects, it is the single most important hand function. It is the motion to repair, retrain, and maintain in the case of an accident.

Inversion—Eversion
(in•**ver**•shun) (ee•**ver**•shun)

Tibia/Fibula–Talus Articulation
(**tib**•ee•ah)(**fib**•yoo•lah)(**tay**•lus)

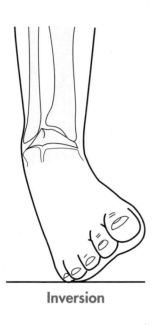

Inversion

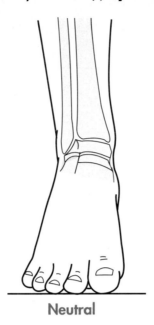

Neutral

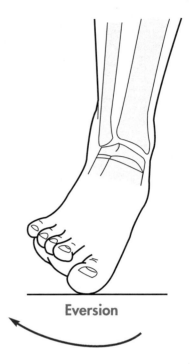

Eversion

Inversion and **eversion** are special movements of the foot. In inversion, the sole of the foot turns inward toward the median line; in eversion, the sole of the foot turns outward or away from the median line.

The ankle joint is the most frequently injured major articulation. **Pott's fracture A** is a result of forced eversion of the foot resulting in fracture of the medial malleolus with the talus shifting laterally, shearing off the lateral malleolus of the fibula superior to the inferior tibiofibular joint. Forced inversion of the foot may cause fracture of the fibular malleolus at the level of the tibial-talus joint **B**.

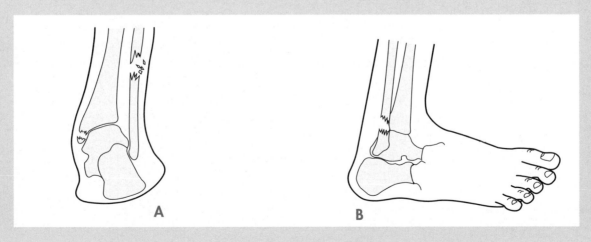

A   B

Temporo-Mandibular (TMJ) Articulation
(**tem**•por•oh)(man•**dib**•yoo•lar)

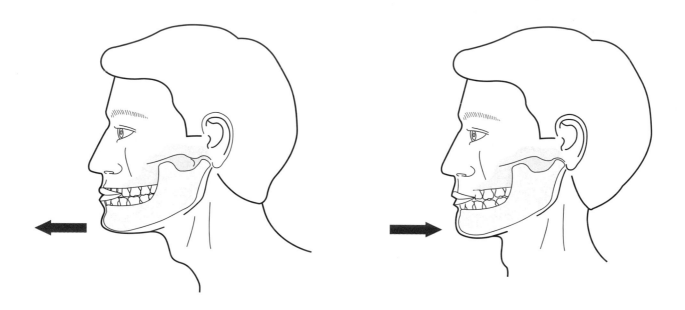

Protraction is movement of a bone anteriorly (forward) in a transverse plane; retraction is movement posteriorly (rearward) along a transverse plane. These are best illustrated in forward and rearward movements of the lower jaw. Similar movements of the shoulder and arms may be made through protraction and retraction of the clavicle and scapula.

Temporo-mandibular joint (TMJ) syndrome is a problem of this joint that has many symptoms, many potential causes, and an equal number of possible treatments. It is characterized by one or more of the following: dull pain around the ear, tenderness of the jaw muscles, clicking or popping noise when opening or closing the mouth, limited or abnormal opening of the mouth, headache, tooth sensitivity, and abnormal wearing of the teeth. The condition may be caused by misalignment of the teeth, missing teeth, poor bite, trauma to the jaw, or arthritis, as well as anxiety, tension, clenching, grinding teeth during sleep (bruxism), or gum chewing.

Temporo-Mandibular (TMJ) Articulation
(**tem**•por•oh)(man•**dib**•yoo•lar)

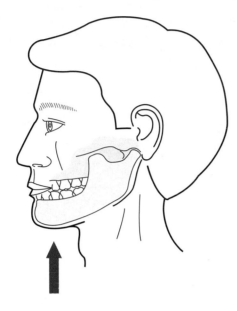 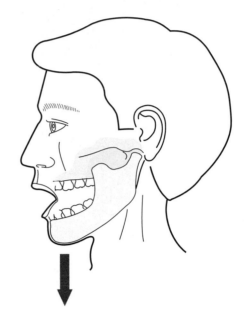

Elevation is a movement of a bone vertically such as upward movement of the clavicles when "shrugging the shoulders" when gesturing "I don't know." Elevating the lower jaw closes the mouth. **Depression** is the opposite motion, lowering the clavicles or lowering the lower jaw (opening the mouth).

If Humans Were Built to Last

by S. Jay Olshansky, Bruce A. Carnes and Robert N. Butler

BIGGER EARS

REWIRED EYES

CURVED NECK

PERSON DESIGNED FOR A HEALTHY OLD AGE
might possess the features highlighted here, along with
countless other external and internal adjustments.

FORWARD-TILTING
UPPER TORSO

SHORTER LIMBS
AND STATURE

EXTRA PADDING
AROUND JOINTS

REVERSED
KNEE JOINT

100 YEARS YOUNG

WALK THIS WAY

A number of the debilitating and even some of the fatal disorders of aging stem in part from bipedal locomotion and an upright posture—ironically, the same features that have enabled the human species to flourish. Every step we take places extraordinary pressure on our feet, ankles, knees and back—structures that support the weight of the whole body above them. Over the course of just a single day, disks in the lower back are subjected to pressures equivalent to several tons per square inch. Over a lifetime, all this pressure takes its toll, as does repetitive use

Flaws

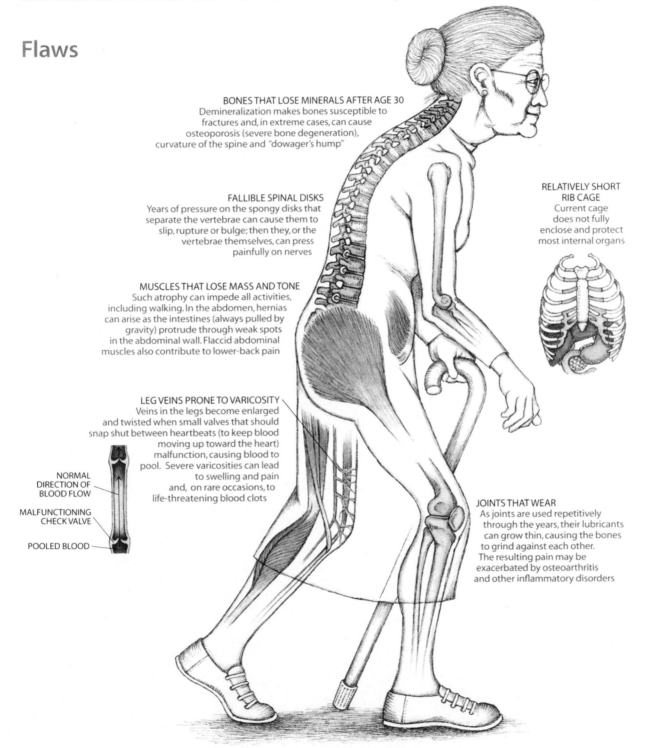

BONES THAT LOSE MINERALS AFTER AGE 30
Demineralization makes bones susceptible to fractures and, in extreme cases, can cause osteoporosis (severe bone degeneration), curvature of the spine and "dowager's hump"

FALLIBLE SPINAL DISKS
Years of pressure on the spongy disks that separate the vertebrae can cause them to slip, rupture or bulge; then they, or the vertebrae themselves, can press painfully on nerves

MUSCLES THAT LOSE MASS AND TONE
Such atrophy can impede all activities, including walking. In the abdomen, hernias can arise as the intestines (always pulled by gravity) protrude through weak spots in the abdominal wall. Flaccid abdominal muscles also contribute to lower-back pain

LEG VEINS PRONE TO VARICOSITY
Veins in the legs become enlarged and twisted when small valves that should snap shut between heartbeats (to keep blood moving up toward the heart) malfunction, causing blood to pool. Severe varicosities can lead to swelling and pain and, on rare occasions, to life-threatening blood clots

NORMAL DIRECTION OF BLOOD FLOW

MALFUNCTIONING CHECK VALVE

POOLED BLOOD

RELATIVELY SHORT RIB CAGE
Current cage does not fully enclose and protect most internal organs

JOINTS THAT WEAR
As joints are used repetitively through the years, their lubricants can grow thin, causing the bones to grind against each other. The resulting pain may be exacerbated by osteoarthritis and other inflammatory disorders

of our joints and the constant tugging of gravity on our tissues.

Although gravity tends to bring us down in the end, we do possess some features that combat its ever present pull. For instance, an intricate network of tendons helps to tether our organs to the spine, keeping them from slumping down and crushing one another.

But these anatomical fixes—like the body in general—were never meant to work forever. Had longevity and persistent good health been the overarching aim of evolution, arrangements such as those depicted below might have become commonplace.

Fixes

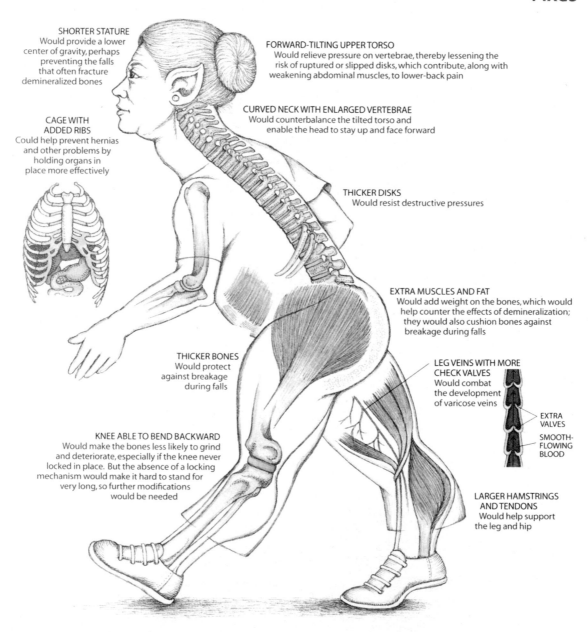

SHORTER STATURE
Would provide a lower center of gravity, perhaps preventing the falls that often fracture demineralized bones

CAGE WITH ADDED RIBS
Could help prevent hernias and other problems by holding organs in place more effectively

FORWARD-TILTING UPPER TORSO
Would relieve pressure on vertebrae, thereby lessening the risk of ruptured or slipped disks, which contribute, along with weakening abdominal muscles, to lower-back pain

CURVED NECK WITH ENLARGED VERTEBRAE
Would counterbalance the tilted torso and enable the head to stay up and face forward

THICKER DISKS
Would resist destructive pressures

EXTRA MUSCLES AND FAT
Would add weight on the bones, which would help counter the effects of demineralization; they would also cushion bones against breakage during falls

THICKER BONES
Would protect against breakage during falls

LEG VEINS WITH MORE CHECK VALVES
Would combat the development of varicose veins

EXTRA VALVES

SMOOTH-FLOWING BLOOD

KNEE ABLE TO BEND BACKWARD
Would make the bones less likely to grind and deteriorate, especially if the knee never locked in place. But the absence of a locking mechanism would make it hard to stand for very long, so further modifications would be needed

LARGER HAMSTRINGS AND TENDONS
Would help support the leg and hip

Muscles of the Face and Head

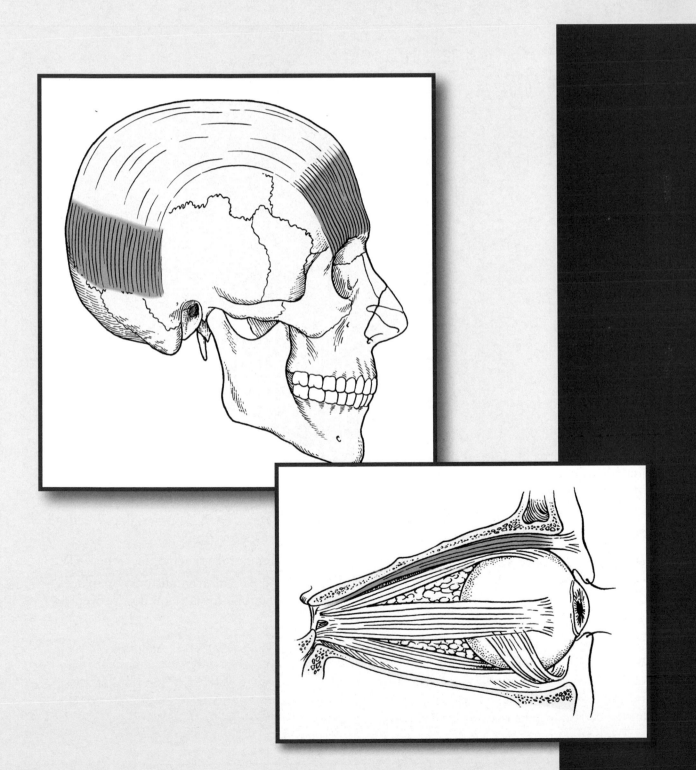

Occipitofrontalis
(ok•**sip**•eh•toh•fron•**tal**•us)

LATERAL VIEW

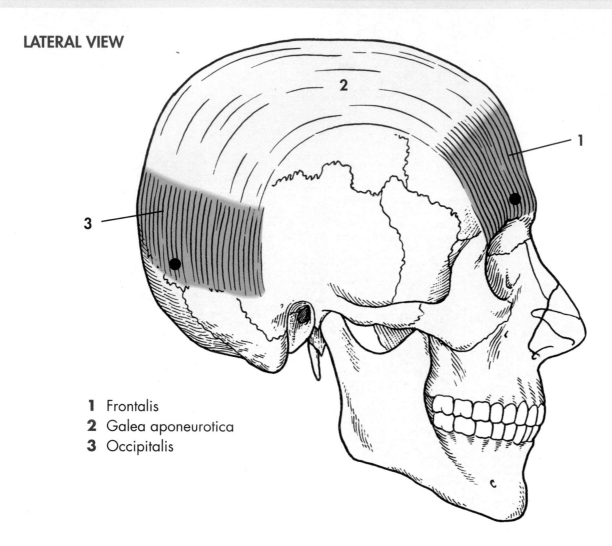

1 Frontalis
2 Galea aponeurotica
3 Occipitalis

Occipital Belly

Origin:	Lateral two-thirds of superior nuchal line of occipital and mastoid process of temporal
Insertion:	Galea aponeurotica covering skull
Action:	Draws back scalp and aids in wrinkling forehead
Innervation:	Posterior auricular branch of facial nerve (VII)

Frontal Belly

Origin:	Galea aponeurotica
Insertion:	Fascia of facial muscles and skin above nose and eyes
Action:	Draws back scalp, wrinkles forehead, raises eyebrows
Innervation:	Temporal branch of facial nerve (VII)
Palpation:	To palpate **frontalis** portion, place hand on forehead and raise eyebrows.

The divisions of the **occipitofrontalis** muscle are connected by a cranial aponeurosis, the galea aponeurotica. The alternate actions of these two muscles pull the scalp forward and backward and assist in wrinkling the forehead and raising the eyebrows. The **trigger point** for the frontalis portion of this muscle is the area above the eyebrow; for the occipitalis portion, it is in the middle just above the nuchal line.

Temporoparietalis
(**tem**•por•oh•pa•rye•eh•**tal**•us)

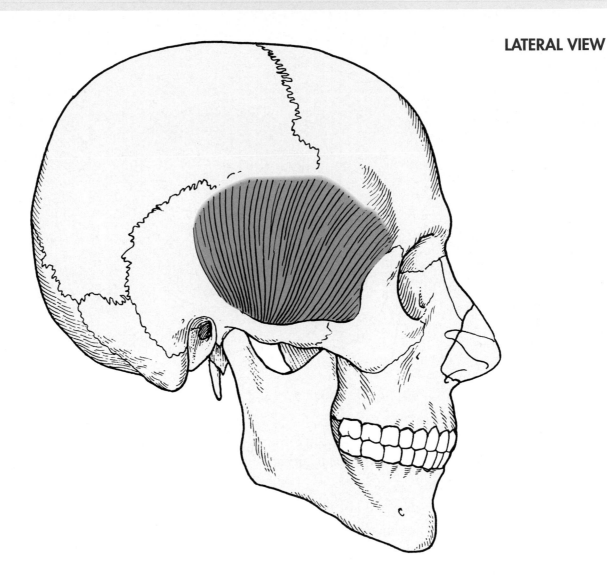

Origin:	Lateral border of galea aponeurotica
Insertion:	Fascia above and cartilage of the auricle
Action:	Raises ears and tightens scalp
Innervation:	Temporal branch of facial nerve (VII)
Palpation:	Contraction is palpable in people who can raise their ears by placing hand 1–2 inches superior and anterior to the ear.

This part of the **epicranius** muscle is superficial to the **temporalis**. It may be very thin or absent in many people, but some people can use it to raise their ears.

Orbicularis Oculi
(or•bik•yoo•**ler**•us)(**ok**•yoo•lye)

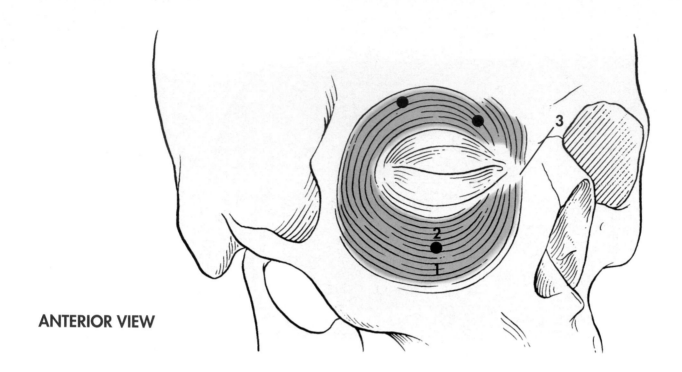

ANTERIOR VIEW

1 Orbital Part

Origin:	Frontal bone and maxilla at medial margin of orbit and palpebral ligament
Insertion:	Same as origin
Action:	Closure of eyelids
Innervation:	Temporal and zygomatic branches of facial nerve (VII)
Palpation:	Palpate superior, lateral, or inferior to eye as the eye is forcefully closed.

2 Palpebral Part

Origin:	Medial palpebral ligament
Insertion:	Lateral palpebral ligament and zygomatic bone
Action:	Closure of eyelid
Innervation:	Temporal and zygomatic branches of facial nerve (VII)

3 Lacrimal Part

Origin:	Lacrimal bone
Insertion:	Lateral palpebral raphe
Action:	Draws lacrimal canals onto surface of eye
Innervation:	Temporal and zygomatic branches of facial nerve (VII)

The **orbicularis oculi** protects the eyes from intense light and injury. The various parts can be activated individually. It produces blinking, winking, and squinting actions and draws the eyebrows inferiorly. Its **trigger points** are in the superior and inferior orbital areas above and below the eye. Its **referred pain pattern** is to the nose.

Levator Palpebrae Superioris
(la•**vay**•ter) (**pal**•pe•bree)(soo•**peer**•ee•**or**•ris)

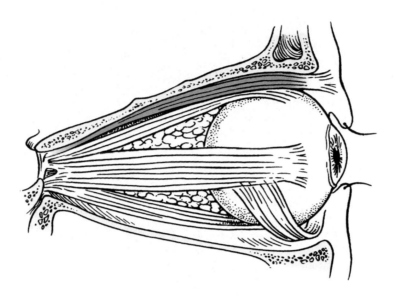

LATERAL VIEW

Origin:	Inferior surface of lesser wing of sphenoid
Insertion:	Tarsal plate of upper eyelid
Action:	Raises upper eyelid or voluntarily opens the eye
Innervation:	Oculomotor nerve (III)
Palpation:	Place finger gently on closed lid, palpate for contraction as eyelid is opened.

"**Drooping eyelid**" or **ptosis** is due to damage to the superior division of the oculomotor nerve, which supplies the levator palpabrae muscle.

Corrugator Supercilii
(kor•ah•gate•ter)(soo•per•sil•e•eye)

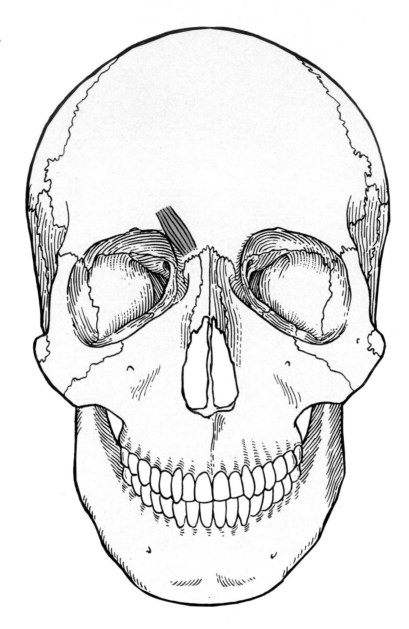

Origin:	Arch of frontal bone above nasal bone
Insertion:	Skin above the middle of the supraorbital margin and above nasal bone
Action:	Draws eyebrows medially and inferiorly
Innervation:	Temporal branch of facial nerve (VII)
Palpation:	Palpate over medial portion of eyebrow when frowning.

This muscle wrinkles the skin of the forehead vertically as in frowning. It also produces the expressions of deep thought, worry, or concern.

Procerus
(pro•**sir**•rus)

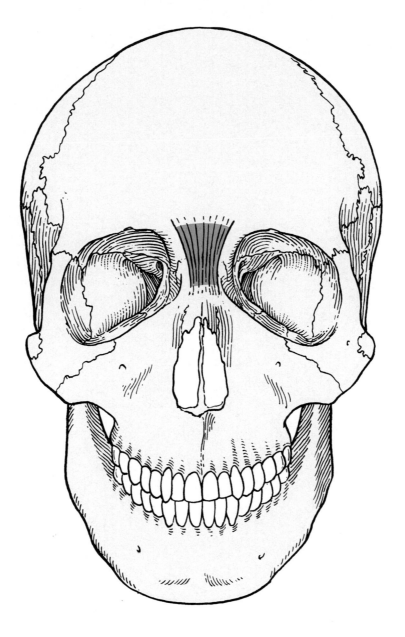

Origin: Fascia over nasal bone and lateral nasal cartilage

Insertion: Skin over the lower forehead between the eyebrows

Action: Draws down medial angles of eyebrows; produces transverse wrinkles over bridge of nose

Innervation: Buccal branches of facial nerve (VII)

Palpation: Palpate on bridge of nose while wrinkling nose.

Nasalis
(nay•**zah**•lis)

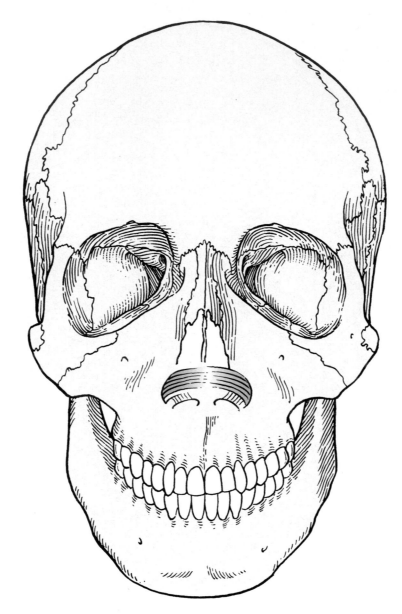

Transverse Part

Origin: Maxilla below infraorbital foramen

Insertion: Muscle from opposite side over bridge of nose

Action: Maintain opening of nares during forceful inspiration

Innervation: Buccal branch of facial nerve (VII)

Palpation: Place fingers on side of nose when breathing deeply.

Alar Part

Origin: Greater alar cartilage of nose

Insertion: Skin at tip of nose

Action: Maintain opening of nares during forceful inspiration

Innervation: Buccal branch of facial nerve (VII)

Both parts of this muscle help flare the nostrils.

Depressor Septi
(de•**press**•ser) (**sept**•eye)

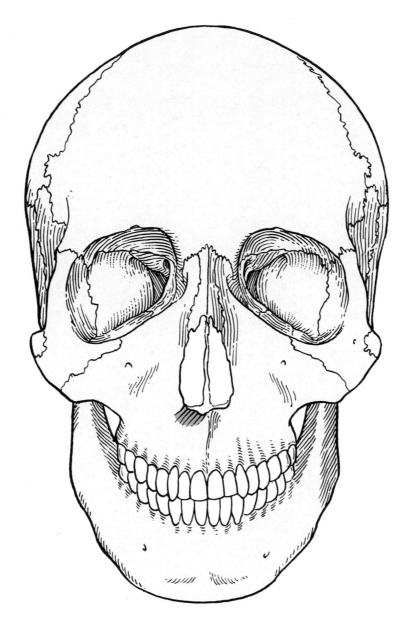

Origin: Maxilla superior to center incisors

Insertion: Nasal septum

Action: Widens nares (nostrils)

Innervation: Buccal branch of facial nerve (VII)

Palpation: Palpate directly inferior to nose as nostrils are constricted.

Orbicularis Oris
(or•bik•u•ler•us) (or•is)

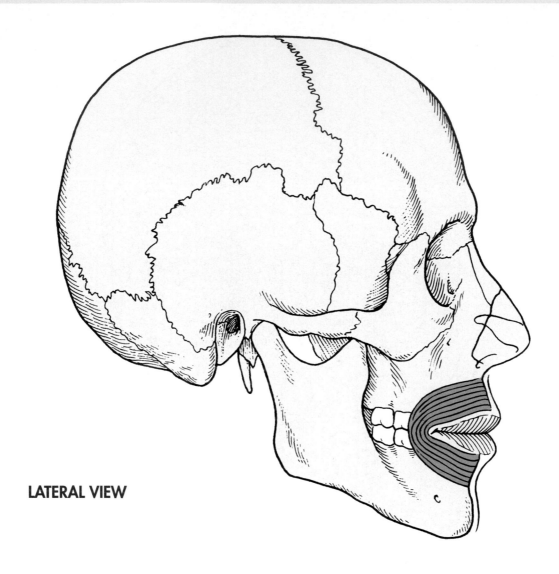

LATERAL VIEW

Origin:	Arises indirectly from the maxilla and mandible; fibers blend with fibers of other facial muscles associated with the lips
Insertion:	Encircles mouth and inserts into muscle and skin at angles of mouth
Action:	Closes and protrudes lips; compresses lips against teeth
Innervation:	Buccal and mandibular branches of facial nerve (VII)
Palpation:	Palpate on tissue of lips as lips are puckered.

This is the "**kissing muscle.**" It is important in whistling and in forming many letters during speech.

Levator Labii Superioris
(la•**vay**•ter) (**lay**•be•eye) (soo•**peer**•ee•**or**•ris)

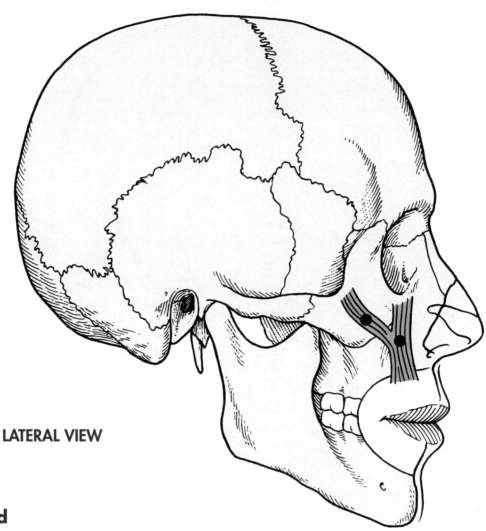

LATERAL VIEW

Angular Head

Origin:	Zygomatic bone and infraorbital margin of maxilla
Insertion:	Skin and orbicularis oris muscle of upper lip
Action:	Elevates upper lip and forms nasolabial furrow
Innervation:	Buccal branch of facial nerve (VII)
Palpation:	Palpate superior to upper lip just lateral to midline as upper lip is elevated.

Infraorbital Head

Origin:	Infraorbital margin of maxilla
Insertion:	Skin and orbicularis oris muscle of upper lip
Action:	Elevates upper lip
Innervation:	Buccal branch of facial nerve (VII)

This thin muscle between the orbicularis oris and the eye margin raises the lip and forms the nasolabial furrow. It is used in "curling" the lip. Its **trigger points** are in the belly of the muscle. Its **referred pain pattern** is below the eye and to the bridge of the nose. It is often associated with sinus pain.

Levator Anguli Oris
(la•**vay**•ter)(**an**•gu•lie)(**or**•is)

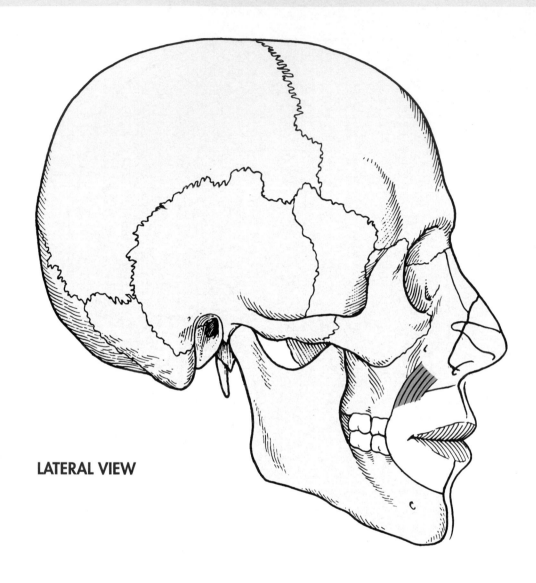

LATERAL VIEW

Origin:	Canine fossa of maxilla
Insertion:	Angle of mouth, blending with fibers of zygomaticus major, depressor anguli oris and orbicularis oris muscles
Action:	Elevates corners of mouth
Innervation:	Buccal branch of facial nerve (VII)
Palpation:	Palpate just superior to angle of mouth while raising lip to show canine tooth.

The **levator anguli oris** acts synergistically with the zygomaticus major in causing an expression of "smiling." It is also important in producing the nasolabial furrow.

Zygomaticus Major—Zygomatic Major
(zye•go•mat•ik•us)(may•jor)

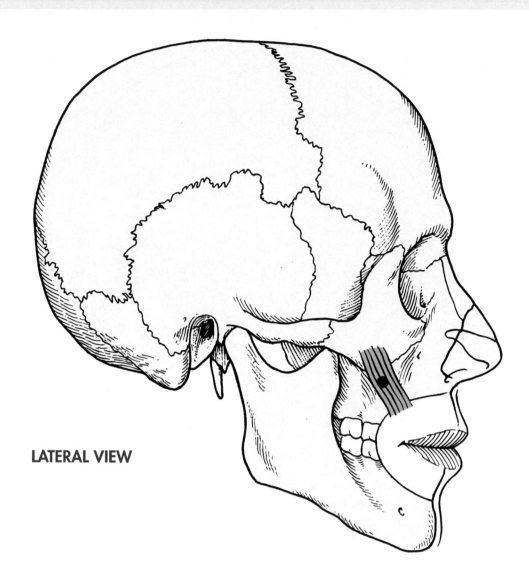

LATERAL VIEW

Origin:	Zygomatic bone
Insertion:	Angle of mouth blending with the levator anguli oris, orbicularis oris, and depressor anguli oris muscles
Action:	Draws angle of mouth upward and outward
Innervation:	Buccal branch of facial nerve (VII)
Palpation:	Palpate superior and lateral to angle of mouth while smiling.

This is the major muscle used in smiling and laughing. The **trigger point** is in the belly of the muscle. Its **referred pain pattern** is below the eye and the side of the nose.

Zygomaticus Minor—Zygomatic Minor

(zye•go•mat•ik•us)

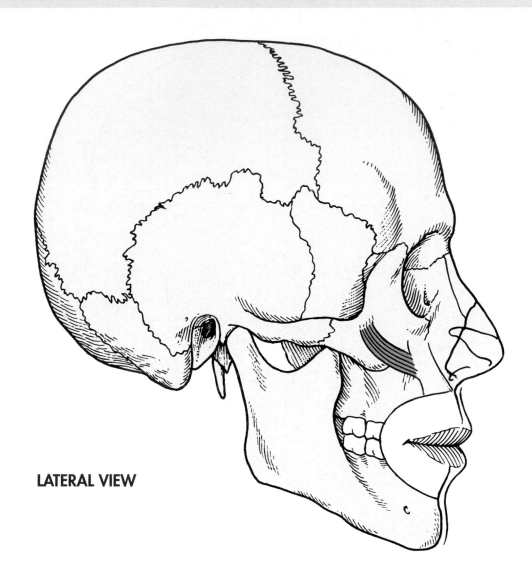

LATERAL VIEW

Origin:	Zygomatic bone
Insertion:	Angle of mouth blending with the levator anguli oris, orbicularis oris, and depressor anguli oris muscles
Action:	Elevates upper lip and produces nasolabial furrow
Innervation:	Buccal branch of facial nerve (VII)
Palpation:	Palpate superior to upper lip about an inch medial to angle of mouth while upper lip is elevated to show teeth.

Risorius
(rye•**sor**•ee•us)

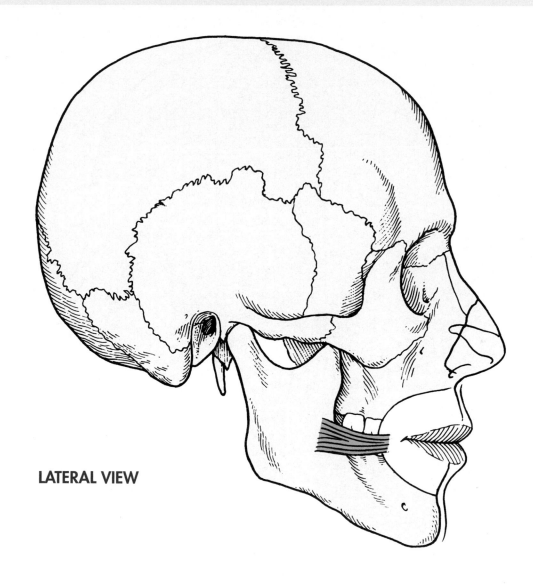

LATERAL VIEW

Origin: Lateral fascia over masseter muscle and parotid gland

Insertion: Skin at angle of the mouth

Action: Draws angle of the mouth laterally

Innervation: Buccal branch of facial nerve (VII)

Palpation: Palpate lateral to angle of mouth when it is drawn laterally.

The point of origin on the **masseter** and its insertion vary; it often inserts a quarter inch inferior to the angle of the mouth. It may be absent. It works **synergistically** with the **zygomaticus major** to tense the mouth and draw the lips into the grinning expression.

Depressor Labii Inferioris
(de•**press**•sor)(**lay**•be•eye)(in•**fear**•ee•or•is)

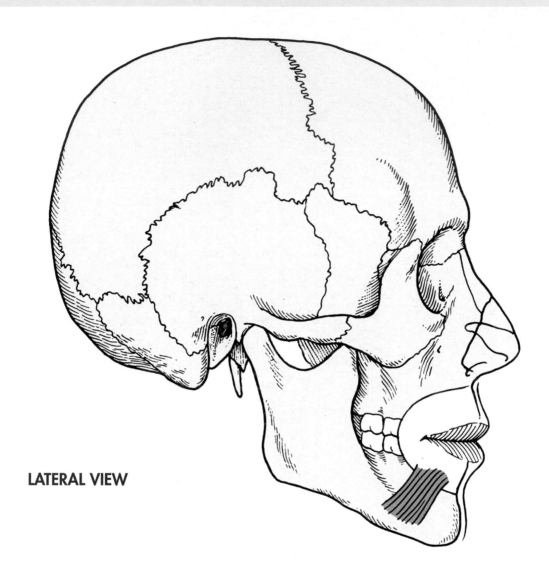

LATERAL VIEW

Origin:	Body of the mandible lateral to the midline, between the mandibular symphysis and the mental foramen
Insertion:	Skin and muscle of lower lip blending with fibers of orbicularis oris
Action:	Draws lower lip inferiorly and laterally during mastication
Innervation:	Mandibular branch of facial nerve (VII)
Palpation:	Palpate inferior and lateral to midline when lower lip is drawn inferiorly.

As this muscle draws the lower lip inferiorly and laterally, it produces the typical expression of a "pout."

Depressor Anguli Oris
(de•**press**•sor) (**an**•gu•lie) (**or**•is)

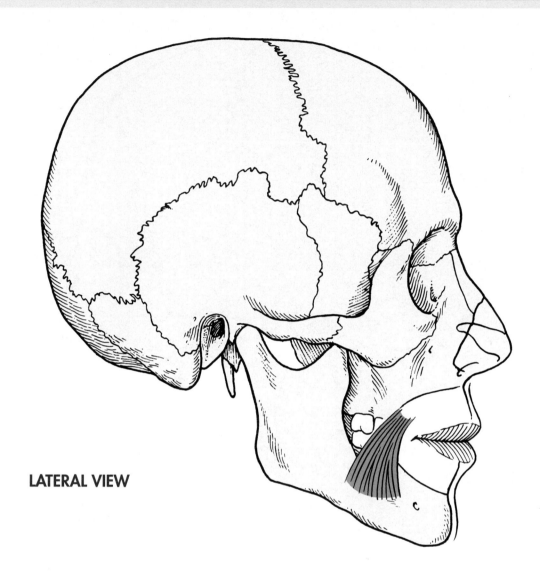

LATERAL VIEW

Origin:	Oblique line of mandible
Insertion:	Skin and muscle at angle of mouth below insertion of zygomaticus
Action:	Draws angle of mouth downward and laterally
Innervation:	Mandibular branch of facial nerve (VII)
Palpation:	Palpate lateral and inferior to angle of mouth when angle of mouth is drawn inferiorly.

This muscle draws the corners of the mouth downward and laterally in opening the mouth, as in the "tragedy mask" grimace. It is also used in frowning, showing disapproval, and expressing the "down-in-the-mouth" grimace.

Mentalis
(men•tal•is)

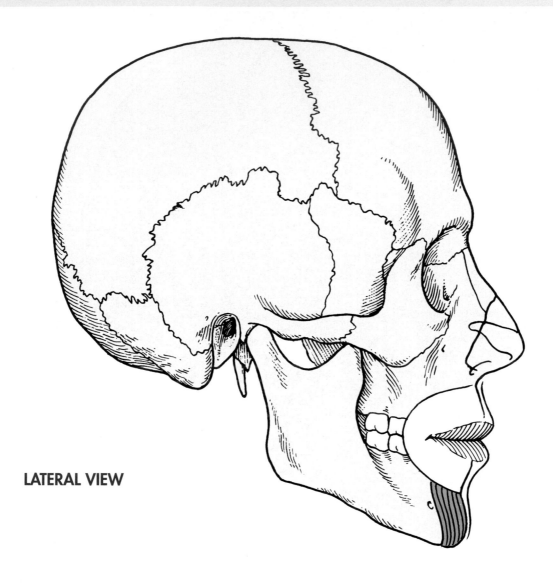

LATERAL VIEW

Origin:	Incisive fossa of mandible
Insertion:	Skin of chin
Action:	Elevates and protrudes lower lip, at same time wrinkling the skin of the chin
Innervation:	Mandibular branch of facial nerve (VII)
Palpation:	Palpate inferior to midline of lower lip while pouting.

This broad muscle forms the muscle mass of the chin. It is used in "pouting."

Buccinator
(buck•si•nay•ter)

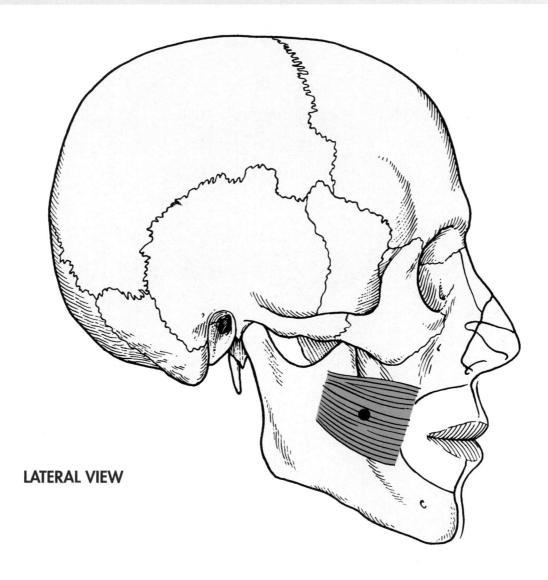

LATERAL VIEW

Origin:	Outer surface of alveolar processes of maxilla and mandible and pterygomandibular raphe
Insertion:	Angle of mouth blending with fibers of the orbicularis oris muscle
Action:	Draws corner of mouth laterally, compresses cheek
Innervation:	Lower buccal branches of the facial nerve (VII)
Palpation:	Palpate on cheek anterior to ramus of mandible while lips are pushed against teeth as in blowing a trumpet.

These muscles are important in compressing the cheeks as in blowing air out of the mouth. They also hold food between the teeth in chewing and cause the cheeks to cave in, producing the sucking action in drinking through a straw. It is well-developed in nursing infants. When it is paralyzed, as in **Bell's palsy**, food accumulates in the oral vestibule. Bell's palsy is a unilateral paralysis of the facial muscles caused by dysfunction of cranial nerve VII. The **trigger point** is in the belly of the muscle. The **referred pain pattern** is to the upper gum.

Temporalis
(tem•po•ral•lis)

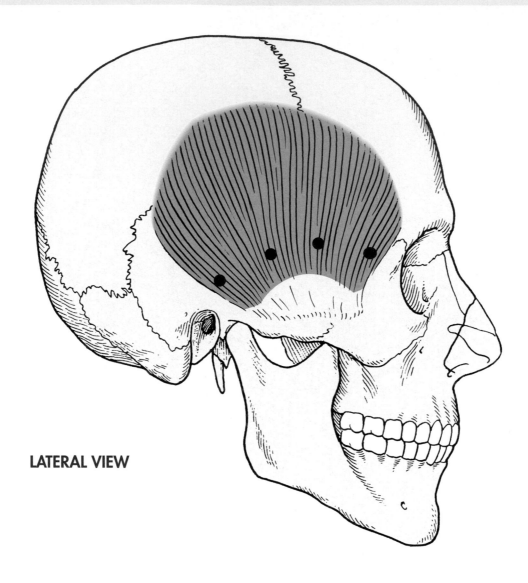

LATERAL VIEW

Origin:	Temporal fossa and temporal fascia
Insertion:	Coronoid process of mandible via a tendon that passes deep to the zygomatic arch
Action:	Elevates and retracts mandible, assists in side to side movement of mandible
Innervation:	Mandibular branch of trigeminal nerve (V)
Palpation:	Palpate temporal fossa as mandible is alternately raised and lowered at the TMJ.

The **temporalis** is a fan-shaped muscle that covers parts of the temporal, frontal, and parietal bone. It maintains the jaw position at rest. The **trigger points** for this muscle are located anteriorly, medially, and posteriorly along the inferior aspect of the muscle near the tendon at the coronoid process of the mandible. The **referred pain patterns** are the temporal region, eyebrow, and upper teeth. **Synergists** are the **masseter** and **medial pterygoid**. Its **antagonist** is the **digastric**.

Masseter
(mass•see•ter)

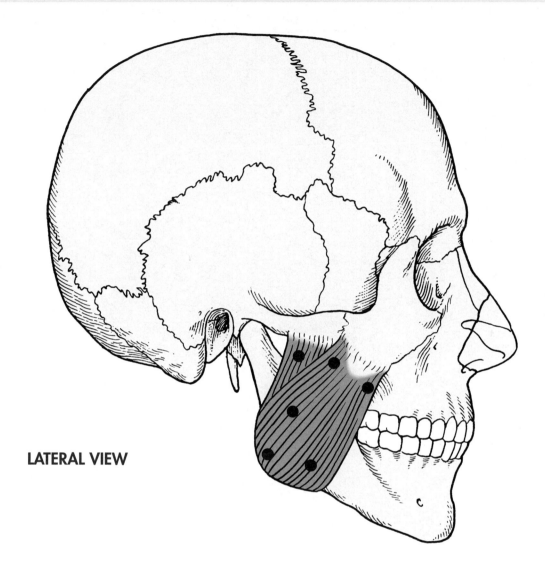

LATERAL VIEW

Origin: Zygomatic process of maxilla and medial and inferior surfaces of zygomatic arch

Insertion: Angle and ramus of the mandible

Action: Elevates mandible and slightly protracts it

Innervation: Mandibular branch of trigeminal nerve (V)

Palpation: Palpate between zygomatic arch and angle of mandible as mandible is raised and lowered.

This cheek muscle bulges as it elevates the lower jaw when you clench your teeth and close the teeth when chewing. The **trigger points** for this muscle are at the tendinous junction near the zygomatic arch and in the belly of the muscle. The **referred pain patterns** are the upper jaw, the ear, and the eyebrow. **Synergists** are the **temporalis** and **medial pterygoid**; its antagonist is the **digastric**.

Pterygoideus Medialis—Medial Pterygoid
(terr•ee•goy•de•us)(mee•de•al•lis)

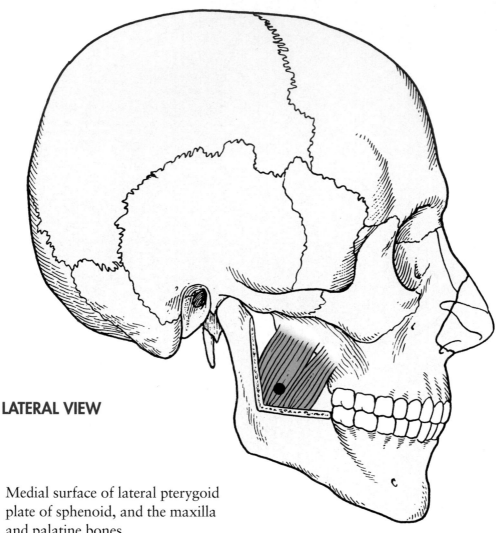

LATERAL VIEW

Origin: Medial surface of lateral pterygoid plate of sphenoid, and the maxilla and palatine bones

Insertion: Posteroinferior aspect of the medial surface of ramus and angle of the mandible

Action: Synergistic with temporalis and masseter in elevation of mandible; it causes protrusion and side-to-side movements of the mandible.

Innervation: Mandibular branch of trigeminal nerve (V)

Palpation: Palpate by placing fingers on angle of mandible and then hook them under onto internal surface while teeth are clenched.

This is a deep, two-headed muscle that runs along the inner surface of the mandible and is largely concealed by that bone. Together with the **lateral pterygoid,** this muscle is primarily responsible for the side-to-side motions involved in chewing and grinding the teeth. The **trigger point** for this muscle is in the belly of the muscle, and it is best accessed from inside the mouth. The **referred pain pattern** is the back of the throat, into the ear. Acts **synergistically** with **lateral pterygoid** in protracting mandible and producing side-to-side grinding movements.

Pterygoideus Lateralis—Lateral Pterygoid

(terr•ee•goy•de•us)(lat•er•al•is)

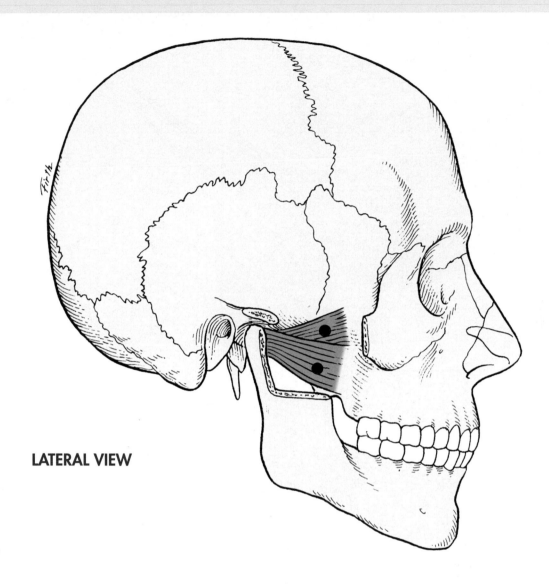

LATERAL VIEW

Origin: A superior head arises from the greater wing of sphenoid bone; an inferior head from the lateral surface of the lateral pterygoid plate of sphenoid

Insertion: Both heads insert on the mandibular condyle and temporomandibular joint capsule

Action: Protrudes, depresses, and moves mandible from side-to-side

Innervation: Mandibular branch of trigeminal nerve (V)

Palpation: Palpate inferior to zygomatic arch between the condyle and coronoid process of mandible while mandible is moving laterally.

The **trigger points** for this muscle are the bellies of both divisions; the **referred pain patterns** are the cheek and the temporomandibular joint. Constant **trigger point** generated tension tends to pull the lower jaw forward and disarticulate the joint. Acts **synergistically** with **medial pterygoid** in protracting the mandible and producing side-to-side grinding movements.

Extrinsic Tongue (Glossal) and Pharyngeal Muscles
(ex•**trin**•sik)   (**gloss**•al)   (fa•**rin**•jee•al)

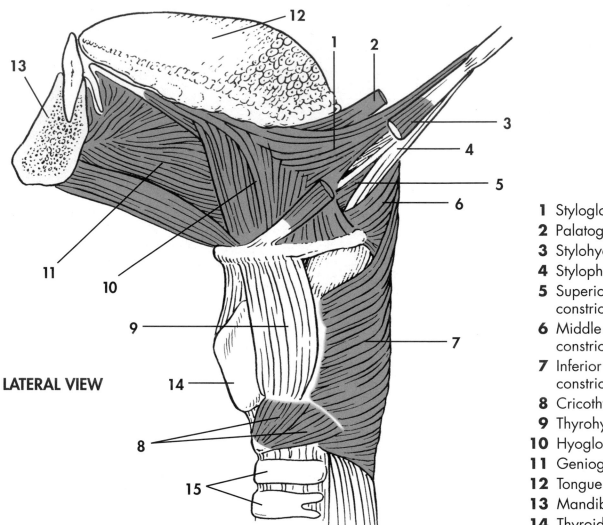

LATERAL VIEW

1 Styloglossus
2 Palatoglossus
3 Stylohyoid
4 Stylopharyngeus
5 Superior pharyngeal constrictor
6 Middle pharyngeal constrictor
7 Inferior pharyngeal constrictor
8 Cricothyroid
9 Thyrohyoid
10 Hyoglossus
11 Genioglossus
12 Tongue
13 Mandibular symphysis
14 Thyroid cartilage
15 Laryngeal cartilages

Intrinsic tongue muscles (those totally within the tongue) change the shape of the tongue, important in chewing and speaking, but do not move it. **Extrinsic tongue muscles** arise from bony structures and project up into the tongue to move it anteriorly, posteriorly, and laterally.

Severe damage of a hypoglossal nerve can cause paralysis of one side of the tongue, causing it to deviate to the paralyzed side during tongue protrusion. Therefore, asking patients to stick out their tongue is a good test of the function of the hypoglossal nerve. When the genioglossal muscle is paralyzed or relaxed during anaesthetization, it tends to fall posteriorly and obstruct the airway, hence the reason for inserting an airtube from the lips to the laryngopharynx.

The muscle cells of the inner muscle layer of the pharynx run superior-inferior. The three circular muscles encircle the inner layer and are arranged one above another, each slightly overlapping the one above. Sequential contraction of these three muscles from superior to inferior produces the force that moves food through the pharynx into the esophagus below.

Extrinsic Tongue (Glossal) and Pharyngeal Muscles

(ex•trin•sik)   (gloss•al)   (fa•rin•jee•al)

Genioglossus
(jee•nee•oh•**gloss**•sus)

Origin: Internal surface of mandible near symphysis

Insertion: Lower portion of tongue and body of hyoid bone

Action: Protracts tongue; can also depress tongue and work with other extrinsic muscles to retract tongue

Innervation: Hypoglossal cranial nerve (XII)

Hyoglossus
(hi•oh•**gloss**•sus)

Origin: Body and greater horn of hyoid bone

Insertion: Lower lateral portion of tongue

Action: Depresses tongue and draws its side downward

Innervation: Hypoglossal cranial nerve (XII)

Palatoglossus
(pa•lat•oh•**gloss**•sus)

Origin: Palatine aponeurosis

Insertion: Side of tongue

Action: Elevates posterior part of tongue and draws soft palate onto tongue

Innervation: Cranial part of spinoaccessory cranial nerve (XI)

Styloglossus
(sty•loh•**gloss**•sus)

Origin: Styloid process of temporal bone

Insertion: Lower lateral portion of tongue

Action: Retracts and elevates tongue

Innervation: Hypoglossal cranial nerve (XII)

Stylopharyngeus
(sty•loh•fa•**rin**•jee•us)

Origin: Medial side of base of styloid process

Insertion: Lateral aspects of pharynx and thyroid cartilage

Action: Elevates larynx and dilates pharynx to help bolus descend

Innervation: Glossopharyngeal nerve (IX)

Superior Pharyngeal Constrictor

Origin: Pterygoid hamulus, pterygomandibular raphe, posterior end of mylohyoid, and side of tongue

Insertion: Median raphe of pharynx and pharyngeal tubercle

Action: Constrict wall of pharynx during swallowing

Innervation: Pharyngeal and superior laryngeal branches of vagus nerve (X)

Middle Pharyngeal Constrictor

Origin: Stylohyoid ligaments and greater and lesser horns of hyoid bone

Insertion: Median raphe of pharynx

Action: Constricts wall of pharynx during swallowing

Innervation: Same as above (X)

Inferior Pharyngeal Constrictor

Origin: Oblique line of thyroid cartilage and side of cricoid cartilage

Insertion: Median raphe of pharynx

Action: Constricts wall of pharynx during swallowing

Innervation: Same as above (X)

Laryngeal Muscles
(lair•**rin**•je•al)

LATERAL VIEW

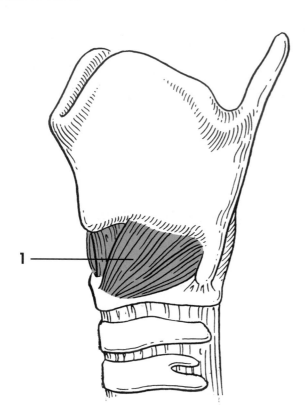

POSTERIOR VIEW

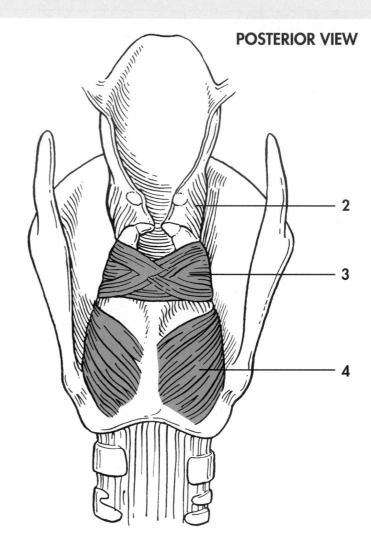

1 Cricothyroideus
2 Thyroarytenoideus
3 Arytenoid
4 Cricoarytenoideus

Arytenoid (ar•e•te•noid)

Origin:	Arytenoid cartilage on one side
Insertion:	Arytenoid cartilage on opposite side
Action:	Close laryngeal aditus by approximating arytenoid cartilages
Innervation:	Laryngeal branch of vagus nerve (X)

Cricoarytenoideus (kri•ko•ar•e•te•noid)

Origin:	Arch of cricoid cartilage
Insertion:	Muscular process of arytenoid cartilage
Action:	Adducts vocal folds
Innervation:	Laryngeal branch of vagus nerve (X)

Cricothyroideus (kri•ko•thi•**roid**•e•us)

Origin:	Anterolateral part of cricoid cartilage
Insertion:	Inferior margin and inferior horn of thyroid cartilage
Action:	Stretches and tenses the vocal folds
Innervation:	Laryngeal branch of vagus nerve (X)

Thyroarytenoideus
(thi•roid•**ar**•e•te•noid•e•us)

Origin:	Posterior surface of thyroid cartilage
Insertion:	Muscular process of arytenoid process
Action:	Relaxes vocal folds
Innervation:	Laryngeal branch of vagus nerve (X)

Muscles of the Soft Palate
(pal•ette)

1 Tensor veli palatini
2 Levator veli palatini
3 Palatopharyngeus
4 Musculus uvulae
5 Salpingopharyngeus
6 Tonsils

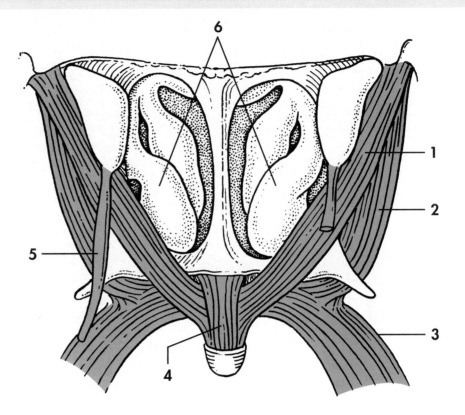

Tensor Veli Palatini
(ten•ser ve•li pal•a•te•ne

Origin: Scaphoid fossa of medial pterygoid plate, spine of sphenoid bone, and cartilage of auditory tube

Insertion: Palatine aponeurosis

Action: Tenses soft palate and opens mouth of auditory tube during swallowing and yawning

Innervation: Medial pterygoid branch of the mandibular branch of trigeminal nerve (V)

Levator Veli Palatini
(le•vat•er ve•li pal•a•te•ne)

Origin: Cartilage of auditory tube and petrous part of temporal bone

Insertion: Palatine aponeurosis

Action: Elevates soft palate during swallowing and yawning

Innervation: Pharyngeal branch of vagus nerve (X)

Palatopharyngeus
(pal•at•o•far•an•je•us)

Origin: Hard palate and palatine aponeurosis

Insertion: Lateral wall of pharynx

Action: Tenses soft palate and pulls walls of pharynx superiorly, anteriorly, and medially during swallowing

Innervation: Cranial part of spinoaccessory (XI) nerve and pharyngeal branch of vagus nerve (X)

Musculus Uvulae (mus•ku•lus yu•vyu•le)

Origin: Posterior nasal spine and palantine aponeurosis

Insertion: Mucosa of uvula

Action: Shortens uvula and pulls it superiorly

Innervation: Cranial part of spinoaccessory nerve (XI) and pharyngeal branch of vagus nerve (X)

Salpingopharyngeus
(sal•pin•jo•far•an•je•us)

Origin: Cartilage of auditory tube

Insertion: Posterior border of thyroid cartilage and side of esophagus

Action: Elevates larynx and pharynx during swallowing and speaking

Innervation: Glossopharyngeal nerve (IX)

Tympanic Cavity Muscles
(tim•pan•ik)

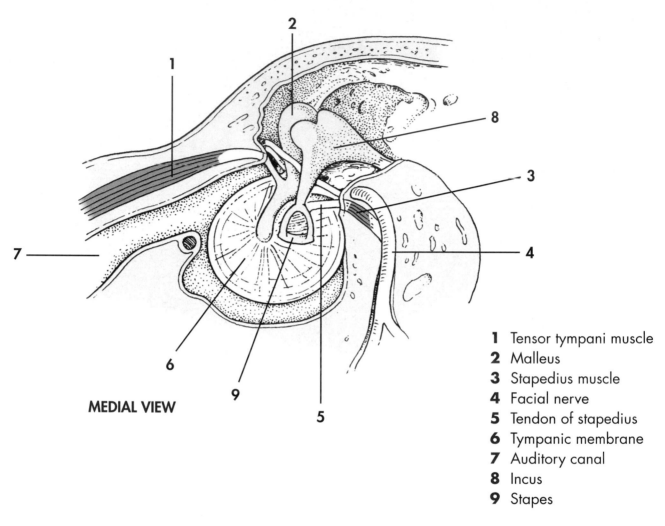

MEDIAL VIEW

1 Tensor tympani muscle
2 Malleus
3 Stapedius muscle
4 Facial nerve
5 Tendon of stapedius
6 Tympanic membrane
7 Auditory canal
8 Incus
9 Stapes

Tensor Tympani (ten•ser tim•pa•ne)

Origin:	Wall of auditory tube
Insertion:	On the malleus ossicle
Action:	(See below)
Innervation:	Trigeminal nerve (V)

Stapedius (sta•pe•de•us)

Origin:	Posterior wall of middle ear cavity
Insertion:	Onto the stapes ossicle
Action:	(See below)
Innervation:	Facial cranial nerve (VII)

The three middle ear ossicles (**malleus, incus, stapes**) articulate with one another by mini synovial joints and span from the tympanic membrane to the bony enclosure of the inner ear. They are suspended by tiny ligaments, and their movement transmits the vibrations of the eardrum to the oval window, which in turn sets fluids of the inner ear into motion, ultimately stimulating the hearing or auditory receptors. The **tensor tympani** and **stapedius** muscles contract reflexively in response to very loud sounds, preventing damage to the hearing receptors. The tensor tympani tenses the eardrum by pulling it medially; the stapedius dampens excessive vibrations to the whole chain of three bones, thus limiting the movement of the stapes in the oval window.

Rectus Extrinsic Eye Muscles
(rek•tus)   (ex•trin•sik)

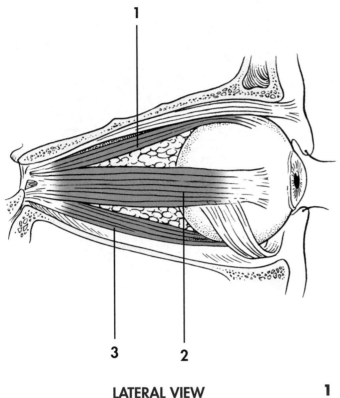

LATERAL VIEW

1 Superior rectus
2 Lateral rectus
3 Inferior rectus
4 Medial rectus

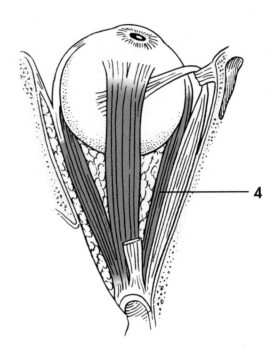

SUPERIOR VIEW

Superior Rectus

Origin:	Annular ring at the back of the bony orbit
Insertion:	Superior surface of anterior sclera
Action:	Elevates eyeball
Innervation:	Oculomotor nerve (III)

Inferior Rectus

Origin:	Annular ring at the back of the bony orbit
Insertion:	Inferior surface of anterior sclera
Action:	Depresses eyeball
Innervation:	Oculomotor nerve (III)

Medial Rectus

Origin:	Annular ring at the back of the bony orbit
Insertion:	Medial surface of the anterior sclera
Action:	Medially rotates eyeball
Innervation:	Oculomotor nerve (III)

Lateral Rectus

Origin:	Annular ring at the back of the bony orbit
Insertion:	Lateral surface of the anterior sclera
Action:	Laterally rotates eyeball
Innervation:	Abducens nerve (VI)

Oblique Extrinsic Eye Muscles
(o•bleek)   (ex•trin•sik)

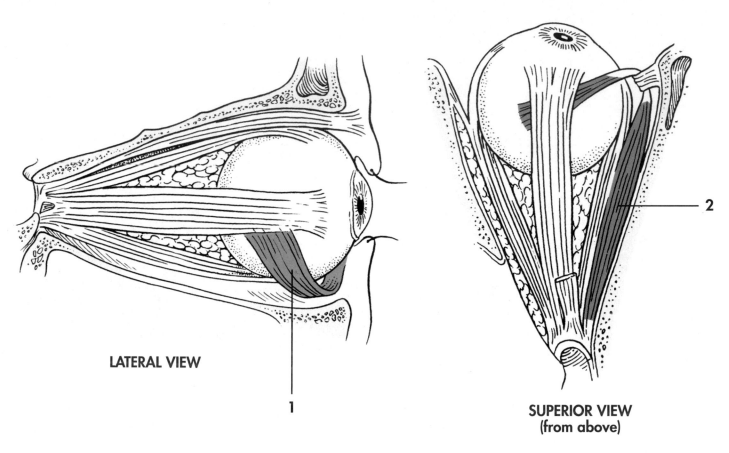

LATERAL VIEW

SUPERIOR VIEW
(from above)

1 Inferior oblique
2 Superior oblique

Inferior Oblique

Origin:	Maxillary bone at the medial inferior corner of the orbit
Insertion:	Lateral surface of the eyeball between the inferior and lateral rectus muscles
Action:	Rotates the eyeball moving it upward and laterally
Innervation:	Oculomotor nerve (III)

Superior Oblique

Origin:	Tendinous ring attached to the bony orbit around the optic foramen
Insertion:	Through a fibrocartilaginous ring, the trochlea, and attaches on the superior surface of the eyeball between the superior and lateral rectus muscles
Action:	Rotates the eyeball moving it downward and laterally
Innervation:	Trochlear nerve (IV)

Muscles of the Neck

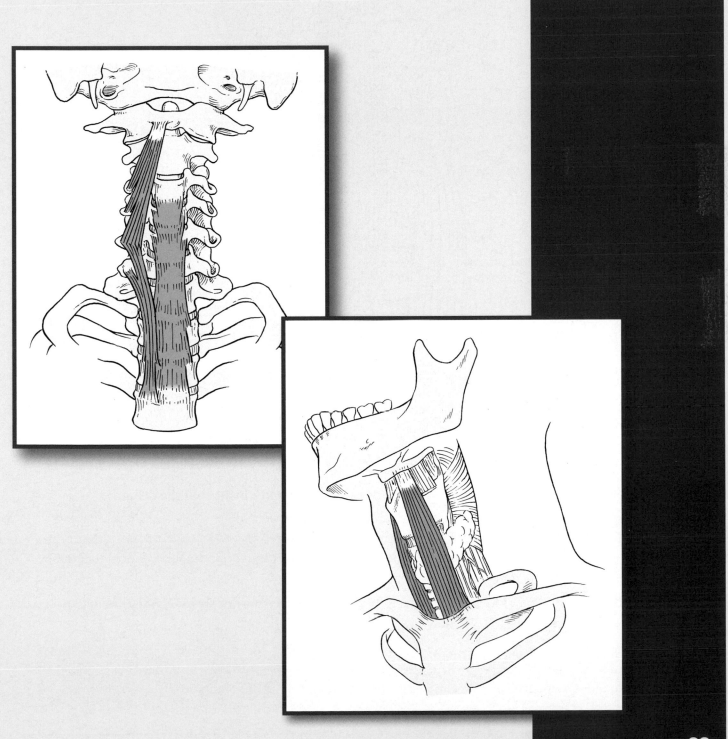

Platysma
(pla•**tiz**•muh)

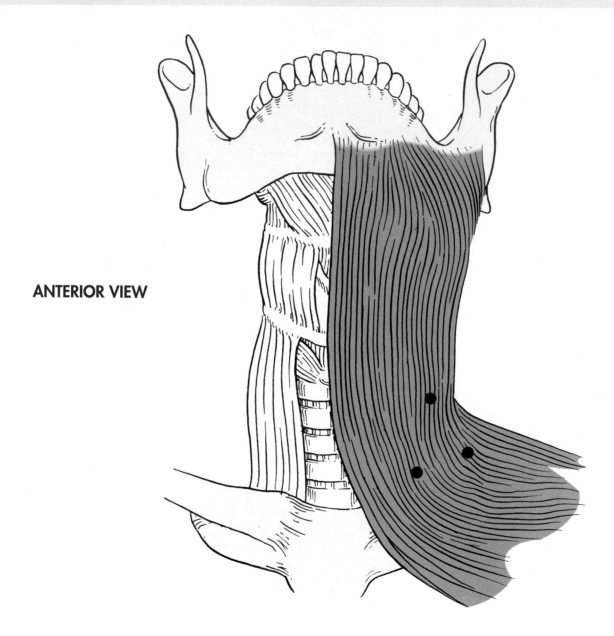

ANTERIOR VIEW

Origin (inferior attachment): Subcutaneous fascia covering the pectoralis major and the deltoid

Insertion (superior attachment): Lower margin of mandible, and subcutaneous fascia and muscles of jaw and mouth

Action: Draws down the lower lip and angle of mouth, tenses skin of neck; helps depress mandible

Innervation: Cervical branch of facial nerve (VII)

Palpation: With hand on anterior neck, forcefully contract **platysma** by depressing and drawing lower lip laterally.

This is the muscle that tenses the neck when shaving. At the TMJ, **synergists** are the **digastric, mylohyoid, and the geniohyoid; antagonists** are the **temporalis, masseter,** and the **medial pterygoid.**

Digastricus—Digastric
(dye•**gas**•tri•kus)

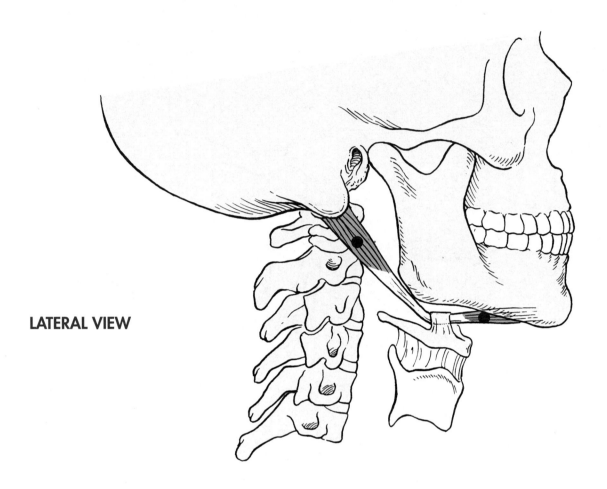

LATERAL VIEW

Origin (superior attachment): **Posterior belly**—between the mastoid and styloid processes of temporal bone

Anterior belly—inner side of inferior margin of mandible near mandibular symphysis

Insertion (inferior attachment): Both bellies insert on the body of the greater cornu of the hyoid bone by a fibrous loop

Action: Acting together, the digastric muscles elevate the hyoid bone and steady it during swallowing and speech; the posterior belly helps open the mouth and depresses the mandible

Innervation: **Anterior belly**—mandibular branch of trigeminal nerve (V)

Posterior belly—cervical branch of facial nerve (VII)

Palpation: Palpate between superior edge of sternocleidomastoid and mandible while mouth is opening.

The **trigger points** for this muscle are in the belly of each division of the muscle. The **referred pain pattern** is in the sternocleidomastoid area and the bottom front teeth. At the TMJ, **synergists** are the **mylohyoid, geniohyoid**, and the **platysma; antagonists** are the **temporalis, masseter**, and the **medial pterygoid**. At the hyoid, **synergists** are the **mylohyoid, geniohyoid**, and the **stylohyoid; antagonists** are the **sternohyoid, thyrohyoid** and the **omohyoid**.

Stylohyoid(eus)
(sty•lo•hi•oid) (ee•us)

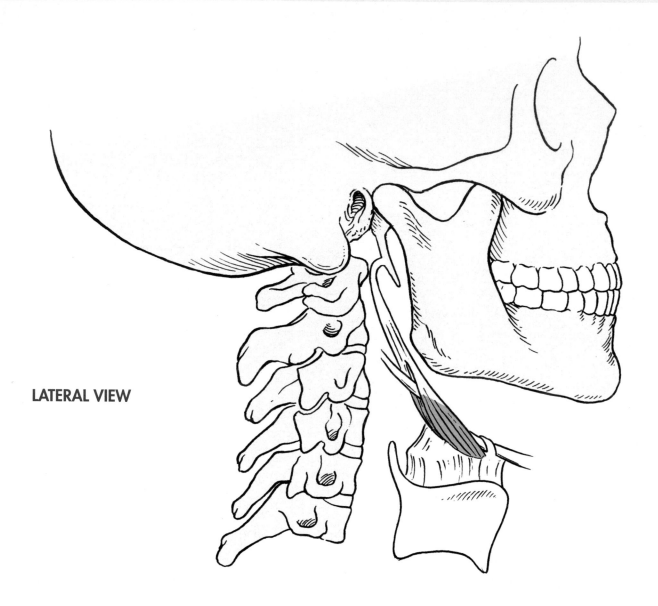

LATERAL VIEW

Origin (superior attachment): Styloid process of temporal bone

Insertion (inferior attachment): Body of hyoid bone at junction of greater cornu

Action: Elevates and retracts the hyoid elongating the floor of the mouth and lifts the tongue during swallowing

Innervation: Facial nerve (VII)

Palpation: Palpate by placing fingers between sternocleidomastoid and posterior ramus of mandible as you swallow.

Synergists are the **digastric, mylohyoid,** and the **geniohyoid; antagonists** are the **sternohyoid, thyrohyoid,** and the **omohyoid.**

Mylohyoid(eus)
(my•lo•hi•oid) (ee•us)

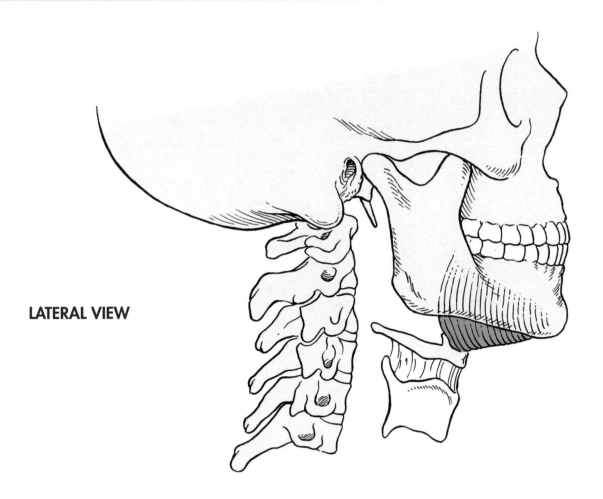

LATERAL VIEW

Origin (superior attachment): Mylohyoid line of mandible

Insertion (inferior attachment): Upper border and median raphe of hyoid bone

Action: Elevates hyoid bone and raises floor of mouth and tongue

Innervation: Mandibular branch of trigeminal nerve (V)

Palpation: Palpate with fingers under chin while placing tip of tongue against roof of mouth.

At the TMJ, **synergists** are the **digastric, geniohyoid,** and the **platysma; antagonists** are the **temporalis, masseter,** and the **medial pterygoid;** at the hyoid, **synergists** are the **digastric, geniohyoid,** and the **stylohyoid; antagonists** are the **sternohyoid, thyrohyoid,** and the **omohyoid.**

Geniohyoid(eus)
(jee•nee•o•hi•oid)(ee•us)

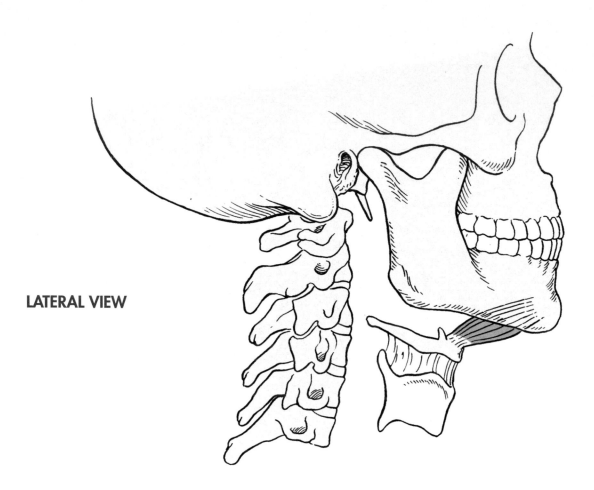

LATERAL VIEW

Origin (superior attachment): Inferior mental spine on inner surface of mandible

Insertion (inferior attachment): Anterior surface of body of hyoid bone

Action: Pulls hyoid bone superiorly and anteriorly shortening the floor of the mouth, it draws the tongue forward

Innervation: First cervical nerve (C1) through the hypoglossal nerve

Palpation: Cannot be palpated separately.

At the TMJ, **synergists** are the **digastric, mylohyoid,** and the **platysma; antagonists** are the **temporalis, masseter,** and the **medial pterygoid;** at the hyoid, **synergists** are the **digastric, mylohyoid,** and the **stylohyoid; antagonists** are the **sternohyoid, thyrohyoid,** and the **omohyoid.**

An Illustrated Atlas of the Skeletal Muscles

Sternohyoid(eus)
(ster•no•hi•oid) (ee•us)

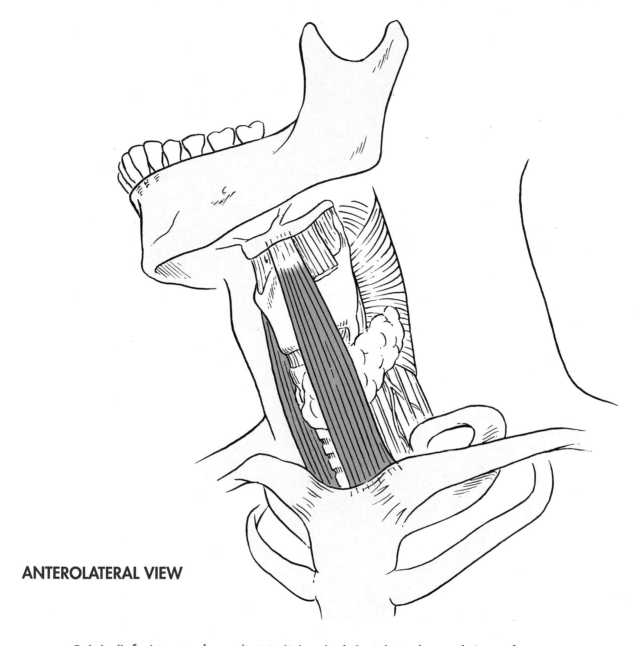

ANTEROLATERAL VIEW

Origin (inferior attachment): Medial end of clavicle and manubrium of sternum

Insertion (superior attachment): Lower margin of body of hyoid bone

Action: Depresses hyoid bone if it has been elevated, as in swallowing

Innervation: Cervical spinal nerves C1–C3 through the ansa cervicalis (slender nerve root in cervical plexus)

Palpation: Palpate by placing fingers just lateral to trachea while swallowing.

Synergists are the **thyrohyoid** and **omohyoid**; antagonists are the **digastric, stylohyoid, mylohyoid,** and **geniohyoid.**

Sternothyroid(eus)

(**ster**•no•**thi**•roid)(ee•us)

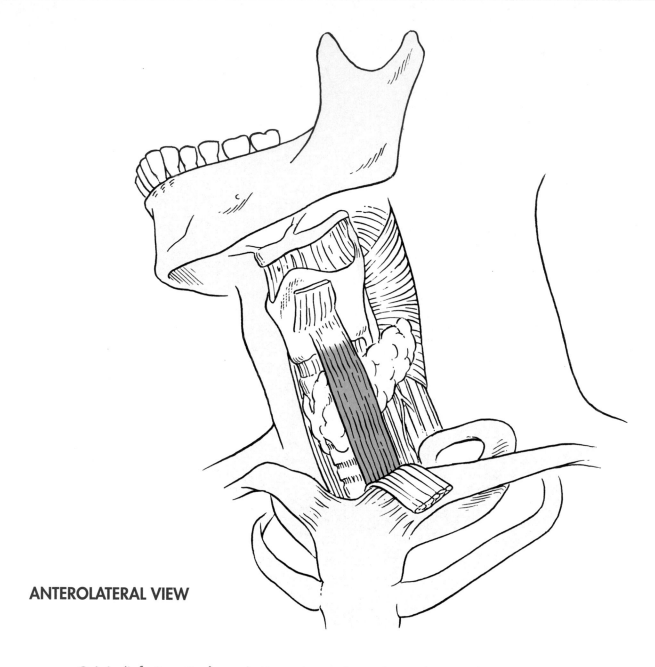

ANTEROLATERAL VIEW

Origin (inferior attachment): Posterior surface of manubrium of sternum

Insertion (superior attachment): Oblique line on lamina of thyroid cartilage

Action: Depresses larynx

Innervation: Ansa cervicalis (C1–C3)

Palpation: Cannot be palpated separately.

Its antagonist is the **thyrohyoid.**

Thyrohyoid(eus)
(thi•ro•hi•oid) (ee•us)

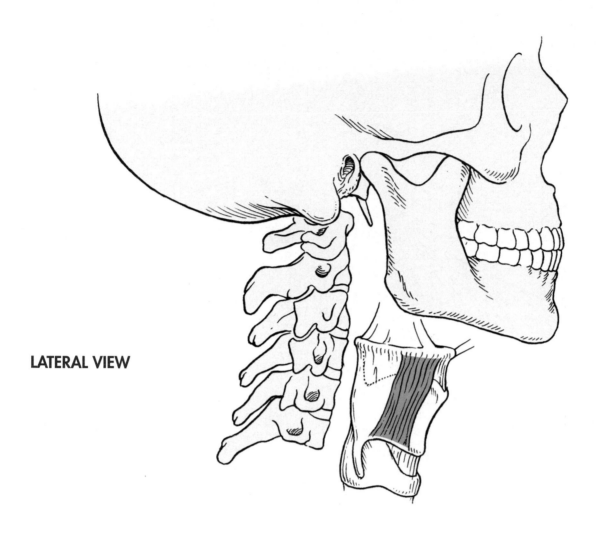

LATERAL VIEW

Origin (inferior attachment): Lamina of the thyroid cartilage at the oblique line

Insertion (superior attachment): Greater cornu of hyoid bone

Action: Depresses hyoid and elevates larynx if hyoid is fixed.

Innervation: First cervical nerve (C1) through the hypoglossal nerve (XII)

Palpation: Cannot be palpated separately.

Synergists are the **sternohyoid** and **omohyoid**; antagonists are the **digastric, stylohyoid, mylohyoid,** and the **geniohyoid.**

Omohyoid(eus)
(o•mo•**hi**•oid) (ee•us)

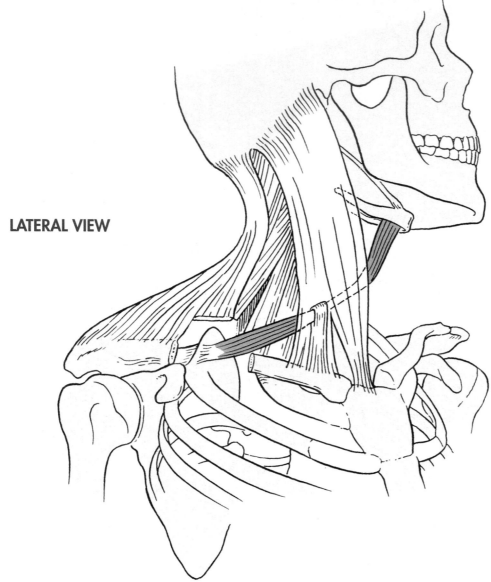

LATERAL VIEW

Origin (inferior attachment): **Superior belly**—arises from tendon of inferior belly near sternocleidomastoid

Inferior belly—superior border of scapula near scapula notch and suprascapula ligaments

Insertion (superior attachment): **Inferior belly**—ends as a tendon (bound to clavicle by central tendon)

Superior belly—inserts on lower border of hyoid bone

Action: Depresses and retracts hyoid bone; retracts larynx

Innervation: Ansa cervicalis (C2–C3)

Palpation: Palpate lateral to sternocleidomastoid just superior to clavicle while depressing mandible against resistance.

Like the digastricus muscle, the **omohyoid** is a straplike muscle with two bellies united by an intermediate tendon. It is lateral to the sternohyoid. **Synergists** are the **sternohyoid** and **thyrohyoid; antagonists** are the **digastric, stylohyoid, mylohyoid** and the **geniohyoid.**

Sternocleidomastoid(eus)
(ster•no•kly•doh•mas•toyd)(ee•us)

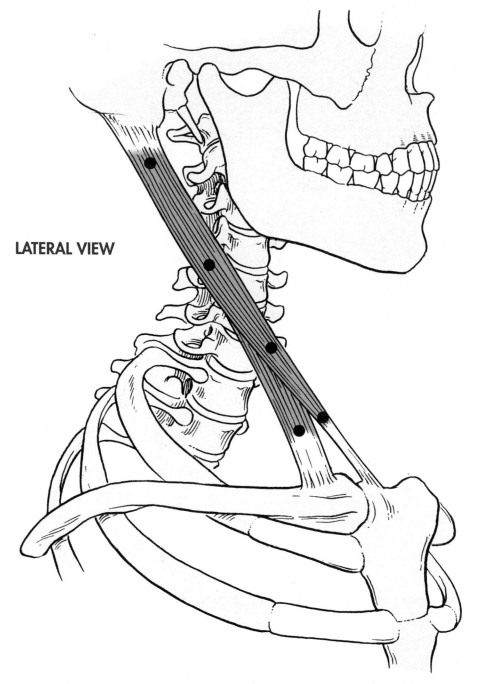

LATERAL VIEW

The fleshy parts of this strap-shaped muscle divide the cervical region into anterior and posterior triangles. It is a key muscular landmark in the neck. Spasms in one of these muscles may cause "wryneck" or torticollis. This can also be a congenital condition where the muscle is injured in a difficult birth, particularly in the breach position. It causes a fixed rotation and tilting of the head due to fibrosis and shortening of the muscles on one side. There are several **trigger points** along the entire length of both heads of the muscle. The **referred pain pattern** is head and face, especially in the occipital region, ear, and forehead. **Synergists** are the **scalene group; antagonists** are the upper divisions of the **erector spinae group.**

Origin (inferior attachment): **Sternal head**—manubrium of sternum

Clavicular head—superior border of medial third of clavicle

Insertion (superior attachment): Mastoid process of temporal and lateral half of superior nuchal line

Action: **Contraction of one side**—bends neck laterally and rotates head to opposite side

Contraction of both sides together—flexes neck; with head fixed it assists in elevating the thorax during forced inspiration

Innervation: Spinal part of spinoaccessory nerve (XI) and branches of cervical spinal nerves (C2–C4)

Palpation: Palpate with hand on anterolateral neck as head is turned and neck flexed.

Longus Capitis
(lon•gus) (kap•eh•tis)

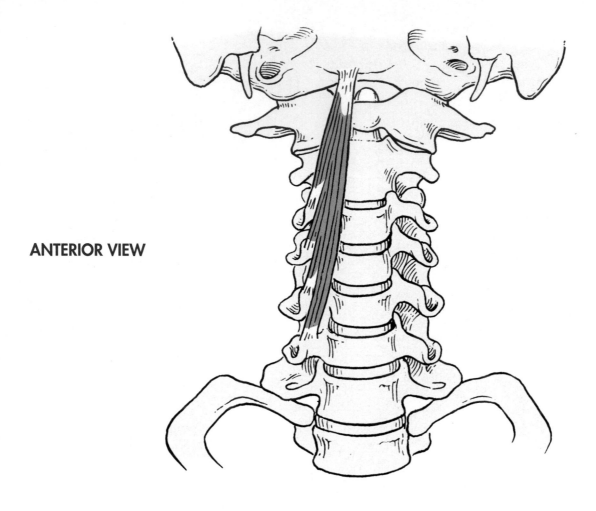

ANTERIOR VIEW

Origin (inferior attachment): Anterior tubercles of the transverse processes of the third through sixth cervical vertebrae

Insertion (superior attachment): Basilar process of occipital bone anterior to foramen magnum

Action: Flexes cervical vertebrae and head

Innervation: C1–C4

Palpation: Deep muscle; cannot be readily palpated.

The superior portion of this muscle may be fused with the **longus colli. Synergists** are the **longus colli, sternocleidomastoid, scalenes muscles, rectus capitis anterior,** and the **platysma;** antagonists are the **rectus capitis posterior major** and **minor,** and the **longissimus capitis.**

Longus Colli
(lon•gus) (koll•lee)

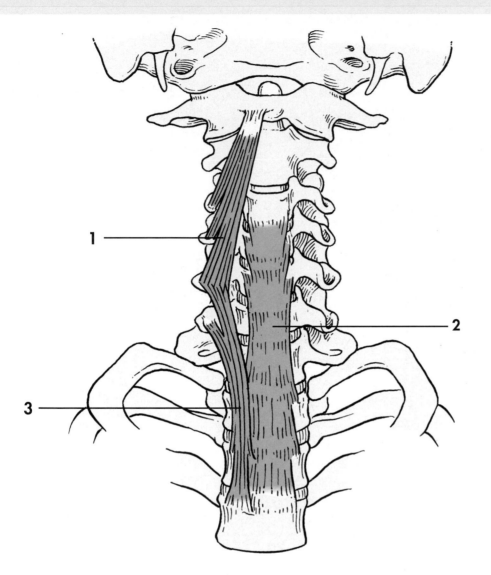

ANTERIOR VIEW

Synergists are the scalenes muscles, the sternocleidomastoid, longus capitis, and the platysma; antagonists are the rectus capitis posterior superior and inferior, the obliquus capitis superior, the longissimus capitis, and the semispinalis capitis.

1 Superior Oblique Part

Origin (inferior attachment): Transverse processes of the third, fourth, and fifth cervical vertebrae

Insertion (superior attachment): Anterior arch of atlas

Palpation: Deep muscle; cannot be readily palpated.

2 Medial Part

Origin (inferior attachment): Anterior surfaces of the bodies of the first three thoracic and lower two cervical vertebrae

Insertion (superior attachment): Anterior surfaces of the third, fourth, and fifth cervical vertebrae

Action: All three parts flex the neck. The superior oblique portion bends it laterally; inferior oblique portion rotates it to the opposite side

Innervation: C2–C7

3 Inferior Oblique Part

Origin (inferior attachment): Anterior surface of the bodies of the first two or three thoracic vertebrae

Insertion (superior attachment): Transverse processes of the fifth and sixth cervical vertebrae

Rectus Capitis Anterior
(rek•tus) (kap•eh•tis)

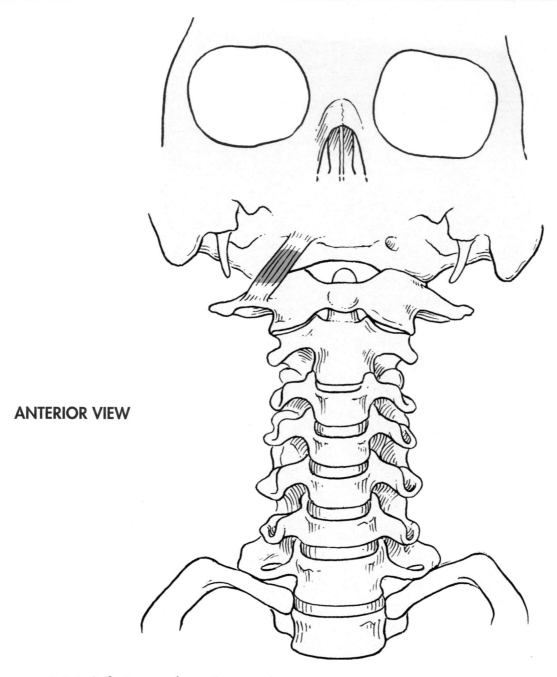

ANTERIOR VIEW

Origin (inferior attachment): Anterior base of transverse process of atlas

Insertion (superior attachment): Occipital bone anterior to the foramen magnum

Action: Flexes head

Innervation: C1, C2

Palpation: Deep muscle; cannot be readily palpated.

Synergists are the **sternocleidomastoid, longus colli,** and **longus capitis; antagonists** are the **rectus capitis posterior major** and **minor, obliquus capitis superior,** and **semispinalis capitis.**

Rectus Capitis Lateralis
(rek•tus) (kap•eh•tis)(lat•er•al•lis)

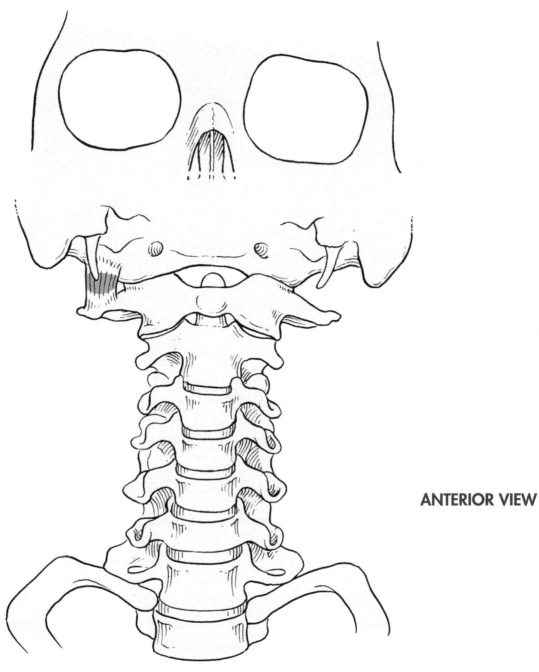

ANTERIOR VIEW

Origin (inferior attachment): Transverse process of atlas

Insertion (superior attachment): Jugular process of occipital bone

Action: Bends head laterally

Innervation: C1, C2

Palpation: Deep muscle; cannot be readily palpated.

Synergists are the **rectus capitis posterior major, obliquus capitis, splenius, sternocleidomastoid,** and **trapezius** muscles on the same side; **antagonists** are the same muscles on the opposite side.

Rectus Capitis Posterior Major
(rek•tus) (kap•eh•tis)

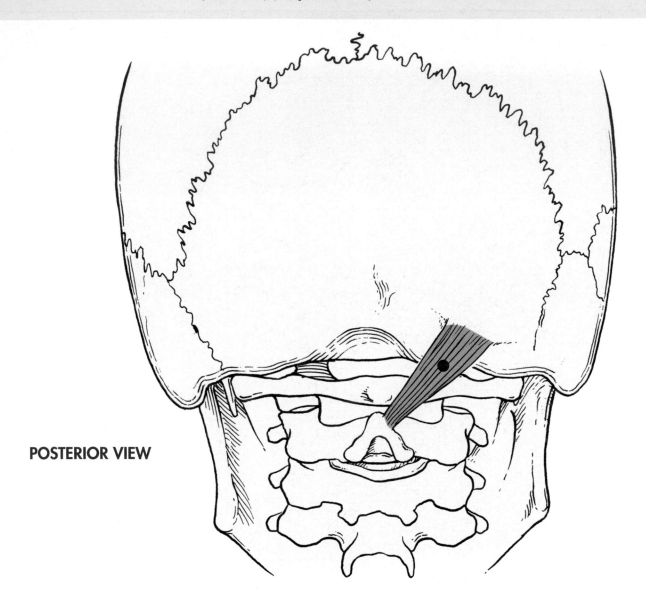

POSTERIOR VIEW

Origin (inferior attachment): Spinous process of axis

Insertion (superior attachment): Lateral portion of inferior nuchal line of occipital bone

Action: Extends and rotates the head toward the same side

Innervation: Dorsal ramus of the suboccipital nerve (C1)

Palpation: Deep muscle; cannot be readily palpated.

The spasmodic contraction of four of the small cervical muscles set off the **referred pain pattern** commonly associated with a **tension headache.** These muscles are the **rectus capitis posterior major, rectus capitis posterior minor, obliquus capitis superior,** and **obliquus capitis inferior.** These muscles are especially important in the movement of the upper two cervical vertebrae. The **trigger point** is in the belly of the muscle. **Synergists** are the **rectus capitis posterior minor, obliquus capitis superior, longissimus capitis,** and **semispinalis capitis;** antagonists are the **rectus capitis anterior, longus colli,** and **longus capitis.**

Rectus Capitis Posterior Minor

(rek•tus) (kap•eh•tis)

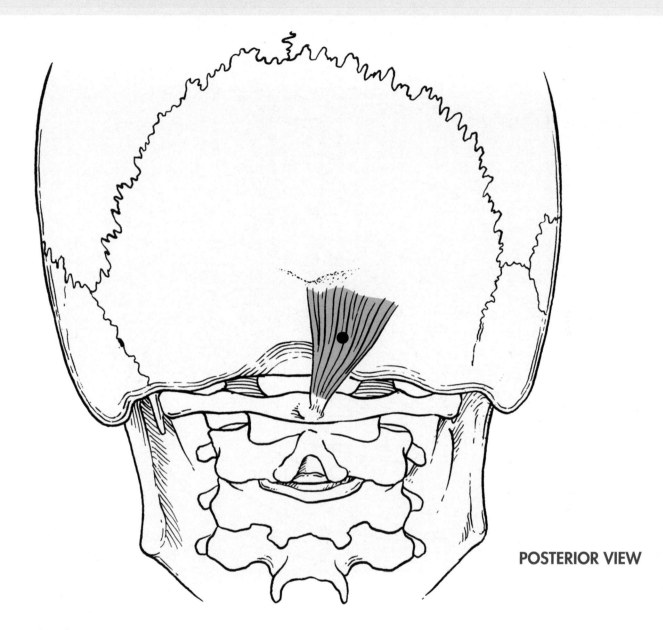

POSTERIOR VIEW

Origin (inferior attachment): Posterior tubercle of atlas

Insertion (superior attachment): Median portion of inferior nuchal line of occipital bone

Action: Extends the head

Innervation: Dorsal ramus of first cervical (suboccipital) nerve (C1)

Palpation: Deep muscle; cannot be readily palpated.

There is difference of opinion about the action of this muscle. It is generally assumed that it extends the head, but **myographic studies** indicate that it does not act in extension, but rather functions as a restraint to flexion and forward movement of the head. The **trigger point** is in the belly of the muscle. **Synergists** are the **rectus capitis posterior major, obliquus capitis superior, longissimus capitis,** and **semispinalis capitis; antagonists** are the **rectus capitis anterior, longus colli,** and **longus capitis.**

Scalenus Anterior—Scalene Anterior

(skay•**lee**•nus)

ANTEROLATERAL VIEW

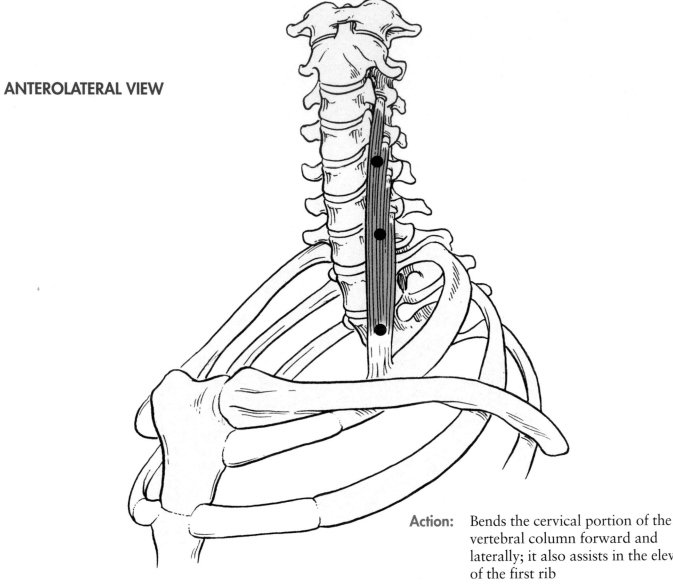

Action: Bends the cervical portion of the vertebral column forward and laterally; it also assists in the elevation of the first rib

Innervation: Ventral rami of the fourth through sixth cervical nerves (C4–C6)

Palpation: Palpate anterior and middle scalenes by placing fingers slightly lateral to the sternocleidomastoid and superior to the clavicle while taking in short deep breaths.

Origin (superior attachment): Anterior tubercle of the transverse processes of the third through sixth cervical vertebrae

Insertion (inferior attachment): Scalene tubercle on the inner border and upper surface of the first rib

A **scalene minimus** may occasionally be present between the first rib and the seventh cervical vertebra. There are multiple **trigger points** along the length of the muscle. The **referred pain pattern** is along the upper and lower arm, lateral side of the hand, and just lateral to the midline in both the anterior and posterior upper thorax. Synergists are the other **scalene muscles, sternocleidomastoid, longus capitis,** and **longus colli; antagonists** are the upper divisions of the **erector spinae muscles, rhomboids,** and **rectus posterior muscles.**

Scalenus Medius—Middle Scalene

(skay•**lee**•nus)(**mee**•de•us)

ANTEROLATERAL VIEW

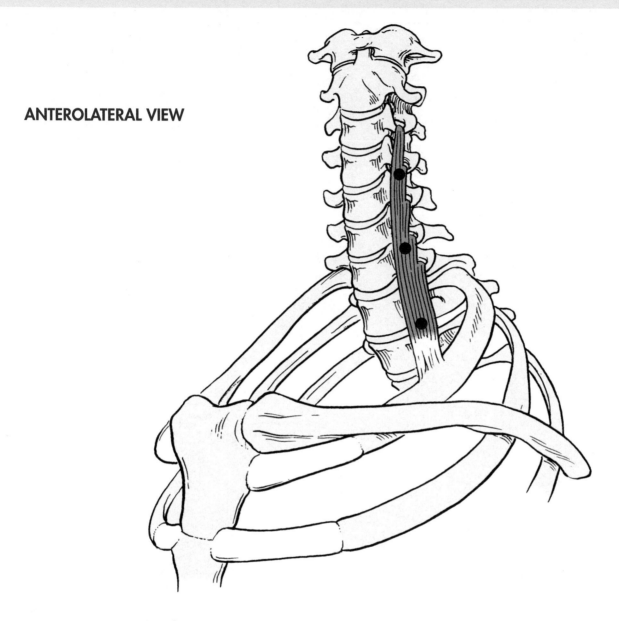

Origin (superior attachment): Front of the posterior tubercles of the transverse processes of the second through seventh cervical vertebrae

Insertion (inferior attachment): Upper surface of the first rib

Action: Acting from above, it helps to raise the first rib; acting from below, it laterally flexes the neck

Innervation: Ventral rami of the fourth through the eighth cervical nerve (C4–C8)

Palpation: Palpated with anterior scalene.

There are multiple **trigger points** along the length of the muscle. The **referred pain pattern** is along the upper and lower arm, lateral side of the hand, and just lateral to the midline in both the anterior and posterior upper thorax. **Synergists** are the other **scalene muscles, sternocleidomastoid, longus capitus,** and **longus colli; antagonists** are the upper divisions of the **erector spinae muscles, rhomboids,** and **rectus posterior muscles.**

Scalenus Posterior—Scalene Posterior
(skay•lee•nus)

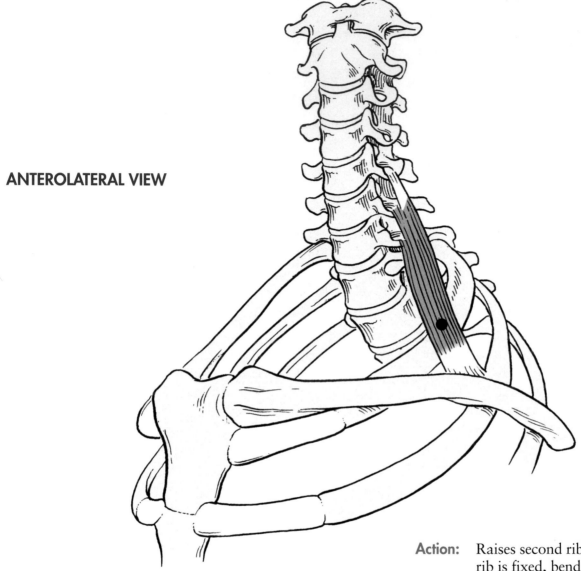

ANTEROLATERAL VIEW

Origin (superior attachment): Posterior tubercles of the transverse processes of the fourth through sixth cervical vertebrae

Insertion (inferior attachment): Outer surface of second rib

Action: Raises second rib; when the second rib is fixed, bends the lower end of the cervical portion of the vertebral column to the same side

Innervation: Ventral rami of the sixth through eighth cervical nerves (C6–C8)

Palpation: Palpate just posterior to where the anterior and middle scalenes are palpated.

The **scalenus group** acts as an accessory muscle of respiration. By raising the first and second ribs, they assist in inspiration. The **trigger point** for each of these muscles is in the belly near the rib attachment points. The **referred pain pattern** is in the pectoral region, the rhomboid region, and the entire length of the arm to the hand. **Synergists** are the other **scalene muscles, sternocleidomastoid, longus capitus,** and **longus colli; antagonists** are the upper divisions of the **erector spinae muscles, rhomboids,** and **rectus posterior muscles.**

Obliquus Capitis Superior
(o•**bleek**•us) (**kap**•eh•tis)

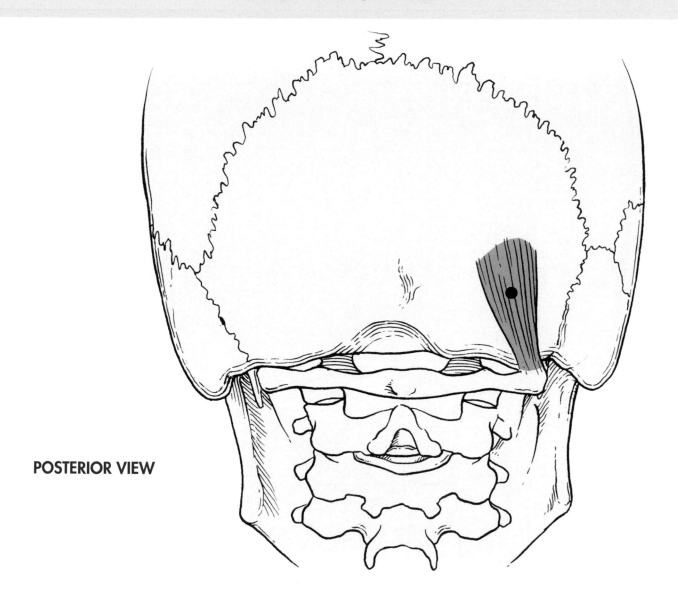

POSTERIOR VIEW

Origin (inferior attachment): Superior surface of transverse process of atlas

Insertion (superior attachment): Occipital bone between inferior and superior nuchal lines

Action: Bends the head backward and laterally to the same side

Innervation: Dorsal ramus of the first cervical (suboccipital) nerve (C1)

Palpation: Deep muscle; cannot be readily palpated.

The **trigger point** is in the belly of the muscle. **Synergists** are the **rectus capitis posterior major** and **minor**, **longissimus capitis**, and **semispinalis capitis**; antagonists are the **rectus capitis anterior**, **longus capitis**, and **longus colli**.

Obliquus Capitis Inferior
(o•**bleek**•us) (**kap**•eh•tis)

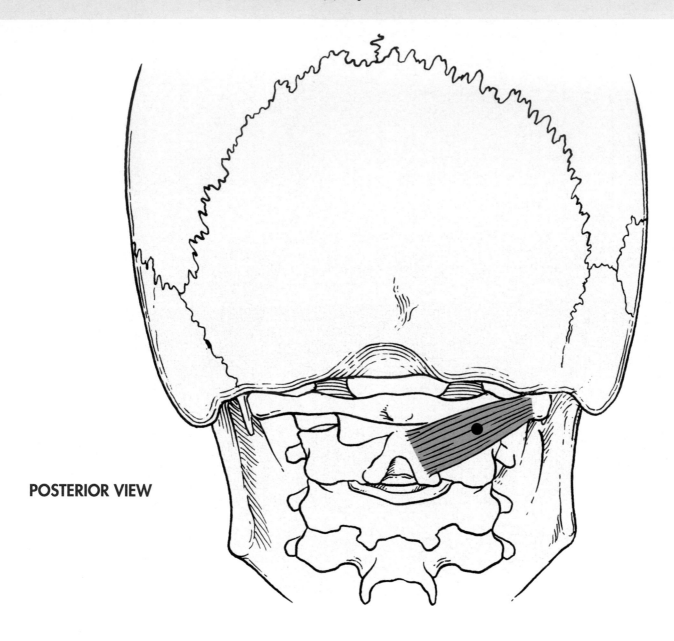

POSTERIOR VIEW

Origin (inferior attachment): Spinous process and upper lamina of axis

Insertion (superior attachment): Transverse process of the atlas

Action: Rotates head

Innervation: Dorsal ramus of the first cervical (suboccipital) nerve (C1)

Palpation: Deep muscle; cannot be readily palpated.

The **trigger point** is in the belly of the muscle. **Synergists** are the **longissimus capitus** and the **rectus capitus posterior major; antagonists** are the **longus colli** and **sternocleidomastoid.**

Muscles of the Torso

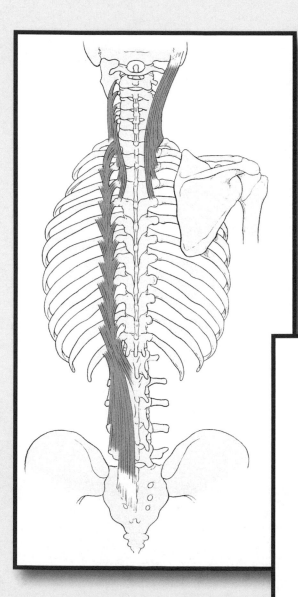

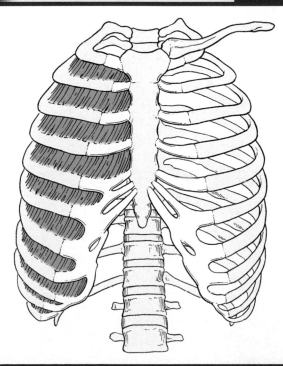

Splenius Capitis
(splee•nee•us**)(kap•**eh•tis**)**

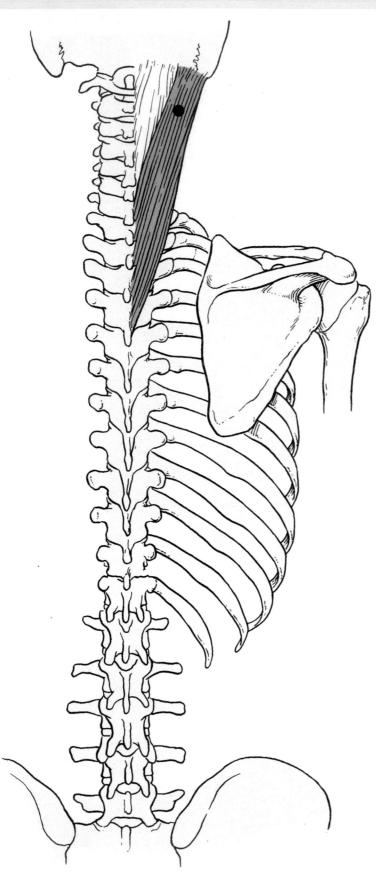

POSTERIOR VIEW

Origin (inferior attachment): Fascia and spinous processes of seventh cervical and first four thoracic vertebrae (C7–T4)

Insertion (superior attachment): Lateral one third of the superior nuchal line and the mastoid process of the temporal bone

Action: Extends and hyperextends the head; contraction of only one side laterally flexes and rotates the head and neck

Innervation: Dorsal rami of the middle cervical nerves (C4–C8)

Palpation: Palpate both **splenius** muscles between posterior edge of sternocleidomastoid and anterior edge of upper trapezius as head and neck are rotating.

The inferior attachment may only go to the third thoracic vertebra. The word **splenius** means bandage. The splenius muscles seem to wrap around the deeper neck muscles. The **trigger point** for the muscle is in the belly close to the head. The **referred pain pattern** is to the top of the head and eye region. **Synergists** are the **splenius cervicis, semispinalis capitis,** and superior portion of the **trapezius; antagonist** is the **sternocleidomastoid.**

Splenius Cervicis
(splee•nee•us)(sir•vih•sis)

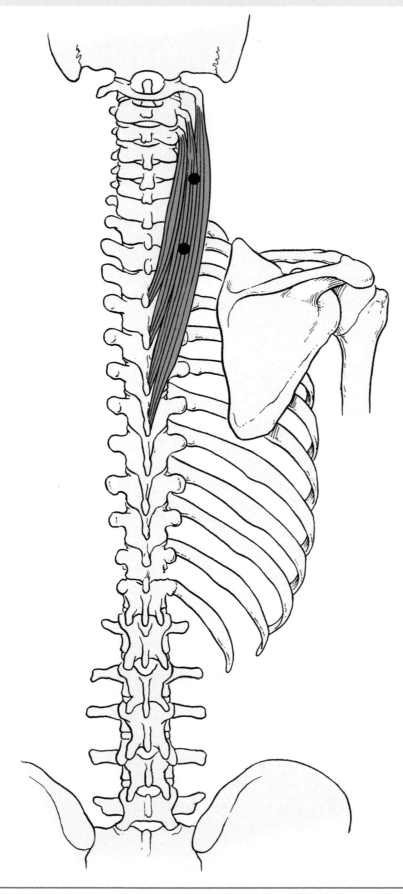

POSTERIOR VIEW

Origin (inferior attachment): Spinous processes of third through sixth thoracic vertebrae (T3–T6)

Insertion (superior attachment): Posterior tubercles of the transverse processes of the first three cervical vertebrae (C1–C3)

Action: The muscles extend and hyperextend the neck, contraction of only one side laterally flexes and rotates the neck and head.

Innervation: Dorsal rami of the lower cervical nerves (C4–C8)

Palpation: Palpate with the **splenius capitis.**

The superior attachment may only be the first two cervical vertebra (C1–C2). The **trigger points** are in the belly of the muscle and near the insertion. **Synergists** are the **splenius capitis, semispinalis capitis,** and superior portion of the **trapezius; antagonists** are the **rectus capitis anterior** and **sternocleidomastoid.**

Erector Spinae Muscles
(e•**rek**•ter) (**spy**•nay)

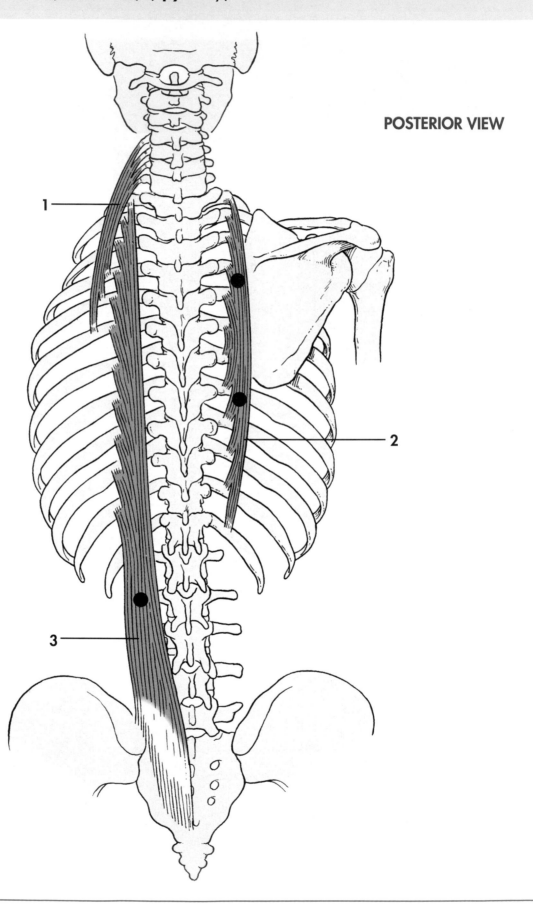

POSTERIOR VIEW

1

2

3

An Illustrated Atlas of the Skeletal Muscles

Erector Spinae Muscles
(e•rek•ter) (spy•nay)

1 Iliocostalis Cervicis
(il•le•oh•cost•tal•lis ser•vis•sis)

Origin (inferior attachment): Angles of the third through sixth ribs

Insertion (superior attachment): Posterior tubercles of the transverse processes of the fourth through seventh cervical vertebrae (C4–C7)

Action: Extension, lateral flexion of the vertebral column

Innervation: Dorsal rami of the lower cervical and thoracic spinal nerves

2 Iliocostalis Thoracis
(il•le•oh•cost•tal•lis tho•ra•sis)

Origin (inferior attachment): Angles of lower six ribs medial to the iliocostalis lumborum

Insertion (superior attachment): Superior border at the angles of the upper six ribs

Action: Extension, lateral flexion of vertebral column

Innervation: Dorsal rami of the thoracic spinal nerves

3 Iliocostalis Lumborum
(il•le•oh•cost•tal•lis lum•bor•um)

Origin (inferior attachment): Medial and lateral sacral crests and medial part of iliac crest

Insertion (superior attachment): Angles of all ribs*

Action: Extension, lateral flexion of vertebral column, lateral movement of pelvis

Innervation: Dorsal rami of the thoracic and lumbar spinal nerves

Palpation: Palpate **erector spinae** muscles as a group lateral to spinous processes of the vertebrae.

---

*Different references indicate insertion varying from lower six to all ribs.

The **erector spinae** muscles are a group of three sets of muscles: the **iliocostalis, longissimus,** and **spinalis.** Together they extend and laterally flex the vertebral column. In the lumbar region they lie deep to the **lumbodorsal fascia** (thoracolumbar) and in the thoracic region they are deep to the **trapezius** and **rhomboideus** muscles. "**Back strain**" is a common **erector spinae** problem in people participating in sports and those who lift heavy weights without proper back support. Back strain indicates some degree of stretching, tearing of muscle fibers or ligaments resulting from excessive extension or rotation of the vertebral column especially in the lumbar region. **Synergists** include the **longissimus, semispinalis, spinalis** groups, and **quadratus lumborum**; its major **antagonist** is the **rectus abdominus.**

Erector Spinae Muscles
(e•**rek**•ter) (**spy**•nay)

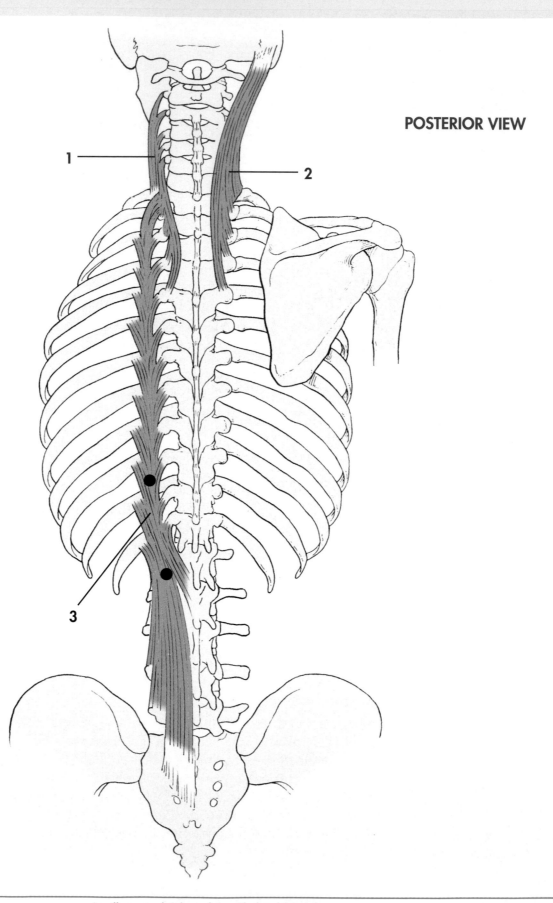

POSTERIOR VIEW

Erector Spinae Muscles
(e•**rek**•ter) (**spy**•nay)

1 Longissimus Cervicis
(lon•**jiss**•i•mus **ser**•vis•sis)

Origin (inferior attachment): Transverse processes of upper five thoracic vertebrae (T1–T5)

Insertion (superior attachment): Posterior tubercles of the transverse processes of the second through sixth cervical vertebrae (C2–C6)

Action: Extension and lateral flexion of vertebral column

Innervation: Dorsal rami of thoracic spinal nerves

2 Longissimus Capitis
(lon•**jiss**•i•mus **cap**•i•tis)

Origin (inferior attachment): Transverse processes of upper five thoracic vertebrae (T1–T5) and articular processes of lower four cervical vertebrae

Insertion (superior attachment): Posterior aspect of mastoid process of temporal bone

Action: Extends and rotates the head

Innervation: Dorsal rami of middle and lower cervical nerves

3 Longissimus Thoracis
(lon•**jiss**•i•mus **tho**•ra•sis)

Origin (inferior attachment): Aponeurosis and transverse processes of lumbar and lower thoracic vertebrae

Insertion (superior attachment): Transverse processes of all thoracic vertebrae and between tubercles and angles of lower ten ribs

Action: Extension and lateral flexion of vertebral column

Innervation: Dorsal rami of thoracic and lumbar spinal nerves

Palpation: Palpate **erector spinae** muscles as a group lateral to spinous processes of the vertebrae.

During full flexion, when bending over, the **erector spinae** muscles are relaxed. Upon standing upright, these muscles are initially inactive and extension is initiated by the hamstring muscles. As a result of this, lifting a load from the bent over position can cause injury to these muscles. **Synergists** to this group include the **iliocostalis, semispinalis,** and **spinalis** groups; its major **antagonist** is the **rectus abdominis.**

Erector Spinae Muscles
(e•**rek**•ter) (**spy**•nay)

POSTERIOR VIEW

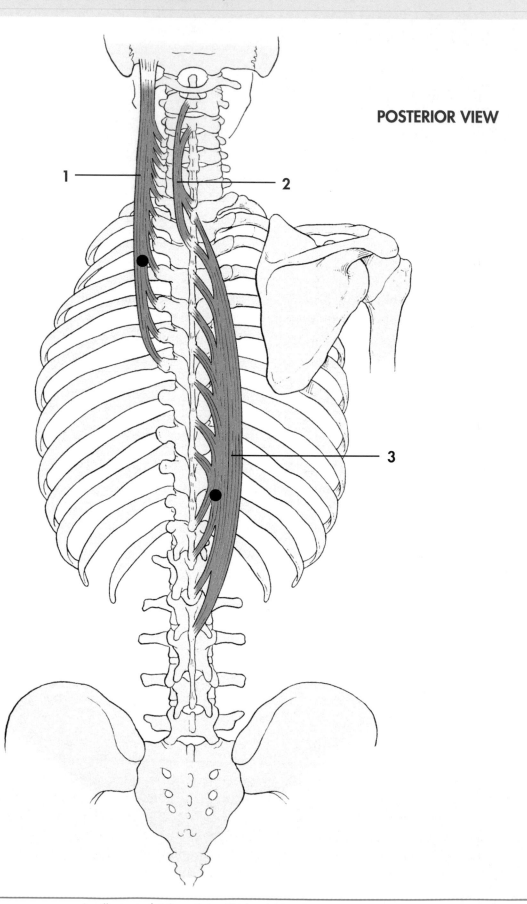

An Illustrated Atlas of the Skeletal Muscles

Erector Spinae Muscles
(e•**rek**•ter) (**spy**•nay)

1 Spinalis Capitis
(spin•**al**•is **cap**•i•tis)

Origin (inferior attachment): Transverse processes of upper seven thoracic (T1–T7) and articular processes of fourth through seventh cervical vertebrae (C4–C7)

Insertion (superior attachment): Between superior and inferior nuchal lines of the occipital bone

Action: Extends the vertebral column

Innervation: Dorsal rami of lower cervical and thoracic spinal nerves

2 Spinalis Cervicis
(spin•**al**•is **ser**•vih•sis)

Origin (inferior attachment): Spinous process of first and second thoracic (T1, T2) and seventh cervical vertebrae (C7)

Insertion (superior attachment): Spinous processes of second and third cervical vertebrae (C2, C3)

Action: Extends the vertebral column

Innervation: Dorsal rami of lower cervical and thoracic spinal nerves

3 Spinalis Thoracis
(spin•**al**•is **tho**•ra•sis)

Origin (inferior attachment): Spinous processes of the lower two thoracic and upper two lumbar vertebrae (T11–L2)

Insertion (superior attachment): Spinous process of upper eight thoracic vertebrae (T1–T8)

Action: Extends the vertebral column

Innervation: Dorsal rami of thoracic and lumbar spinal nerves

Palpation: Palpate **erector spinae** muscles as a group lateral to spinous processes of vertebrae.

The **erector spinae** muscles go into powerful spasms following injury to back structures. **Trigger points** for this group are usually found in the midscapular and lumbar regions. The **referred pain pattern** is to the scapular, lumbar, gluteal, and abdominal regions. Synergists are the **longissimus, semispinalis,** and **iliocostalis** group.

Transversospinalis
(trans•ver•so•spin•al•is)

POSTERIOR VIEW

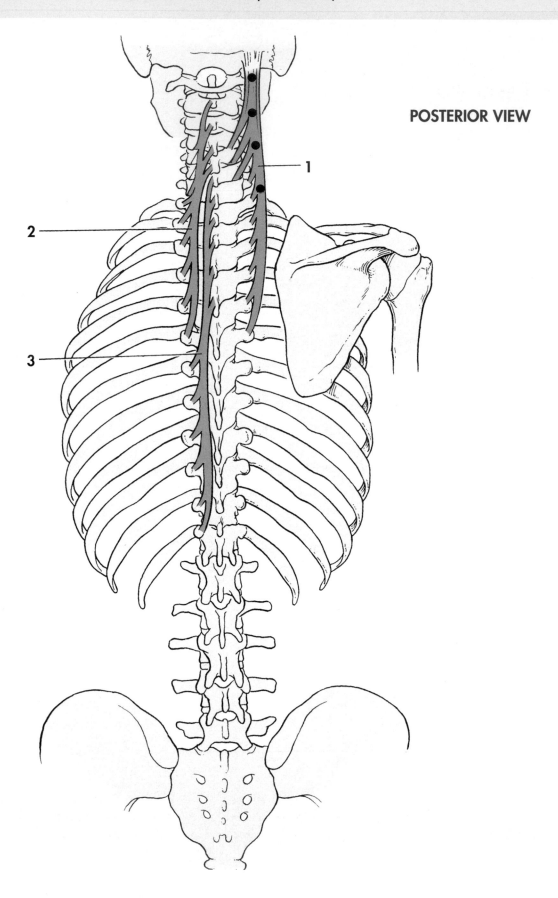

1

2

3

An Illustrated Atlas of the Skeletal Muscles

Transversospinalis
(trans•ver•so•spin•al•is)

1 Semispinalis Capitis
(sem•ee•spin•al•liss **cap**•i•tiss)

Origin (inferior attachment): Transverse process of the upper six thoracic (T1–T6) and seventh cervical (C7) and articular processes of the fourth through six cervical vertebrae (C4–C6)

Insertion (superior attachment): Between the superior and inferior nuchal lines of the occipital bone

Action: Extension of the head and rotation to the opposite side

Innervation: Dorsal rami of the first six cervical spinal nerves (C1–C6)

2 Semispinalis Cervicis
(sem•ee•spin•al•liss **ser**•vih•siss)

Origin (inferior attachment): Transverse processes of upper six thoracic (T1–T6) and articular processes of lower four cervical vertebrae (C4–C7)

Insertion (superior attachment): Spinous process of second through fifth cervical vertebrae (C2–C5)

Action: Extension and rotation of vertebral column

Innervation: Dorsal rami of lower three cervical spinal nerves (C4–C7)

3 Semispinalis Thoracis
(sem•ee•spin•al•is **tho**•ra•sis)

Origin (inferior attachment): Transverse processes of lower six thoracic vertebrae (T7–T12)

Insertion (superior attachment): Spinous processes of the lower two cervical and upper four thoracic vertebrae (C6–T4)

Action: Extension and rotation of vertebral column

Innervation: Dorsal rami of upper six thoracic spinal nerves (T1–T6)

Palpation: Deep muscles; cannot be readily palpated.

The **trigger points** are in the upper portion of the semispinalis capitis muscle. The **referred pain pattern** is to the rear of the head and a band encircling the head from the occipital to the eye. **Synergists** are the **longissimus, iliocostalis,** and **spinalis** group; its **antagonist** is the **rectus abdominis.**

Multifidus
(mul•tif•eh•dus)

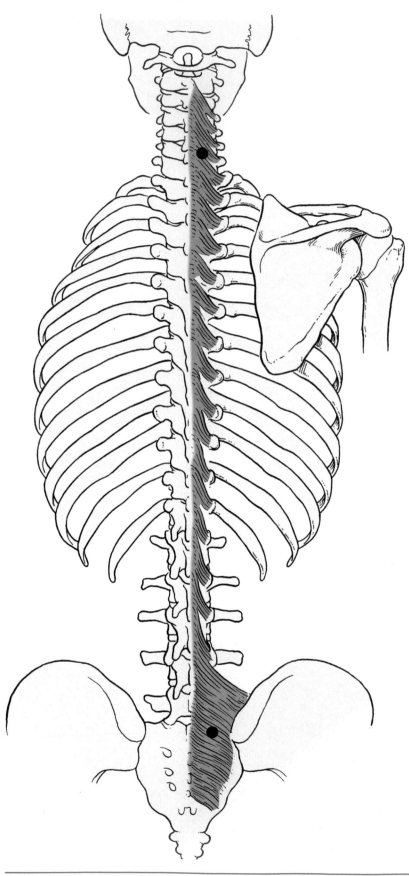

Origin (inferior attachment): Articular processes of the last four cervical, transverse processes of all thoracic, and mammillary processes of lumbar vertebrae, the posterior superior iliac spine, posterior sacroiliac ligaments, and dorsal surface of sacrum adjacent to sacral spinous processes

Insertion (superior attachment): Spinous process of the vertebra above the vertebra of origin

Action: Extend and rotate vertebral column

Innervation: Dorsal rami of spinal nerves

Palpation: Place hands on either side of sacrum while upper part of body is being actively extended.

The **multifidi** are part of the **transversospinalis** group of muscles. They lie deep to the **erector spinae**. This group extends and rotates the spine. The **trigger points** are in the belly of the muscle in the lumbosacral region and in the cervical region. The **referred pain pattern** may feel like it is in the spine itself because tension in these small diagonal muscles may pull one or more vertebrae out of line to one side, pressing nerves and producing additional pain. **Synergists** are the **rotatores, interspinales,** and **intertransversarii** groups.

Rotatores
(ro•tah•**tor**•eez)

POSTERIOR VIEW

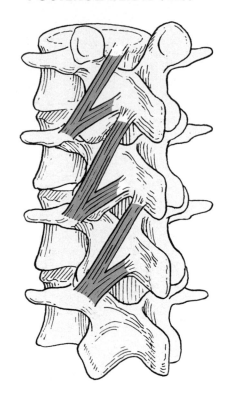

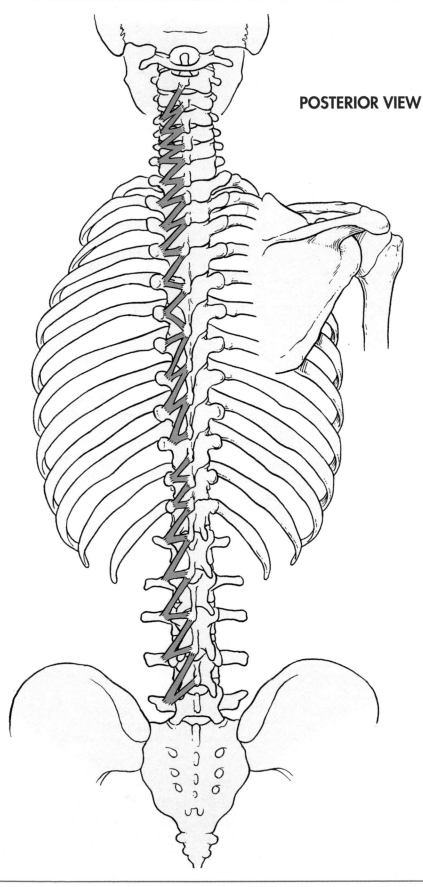

Origin (inferior attachment): Transverse processes of each vertebrae

Insertion (superior attachment):

Short head—base of spinous process of next vertebrae above

Long head—base of spinous process of second vertebra above

Action: Extend and rotate the vertebral column

Innervation: Dorsal rami of spinal nerves

Palpation: Deep muscles; cannot be palpated.

Synergists are the **intertransversarii**, **interspinales**, and **multifidus**; their **antagonist** is the **rectus abdominis**.

Interspinales

(in•ter•spy•nay•les)

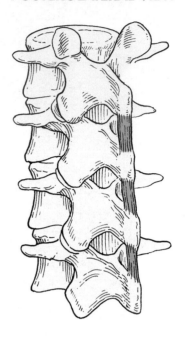

POSTEROLATERAL VIEW

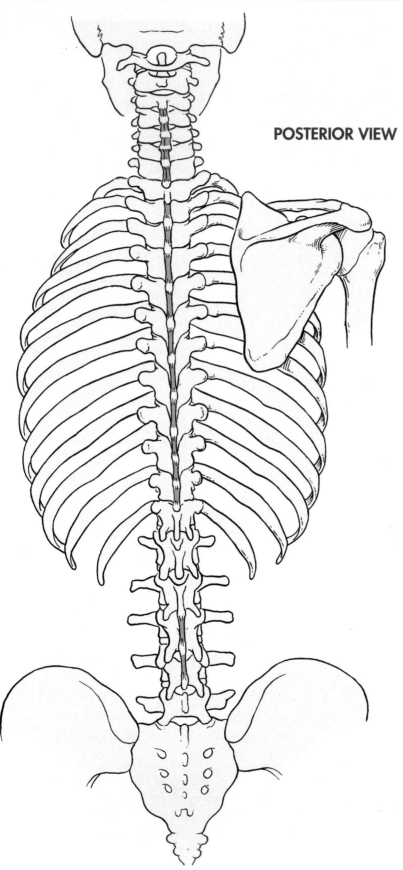

POSTERIOR VIEW

Origin (inferior attachment): Cervical region—spinous processes of third to seventh cervical vertebrae (C3–C7)

Thoracic region—spinous processes of second to twelfth thoracic vertebrae (T2–T12)

Lumbar region—spinous processes of second to fifth lumbar vertebrae (L2–L5)

Insertion (superior attachment): Spinous process of next superior vertebra to the vertebra of origin

Action: Extend the vertebral column

Innervation: Posterior primary rami of spinal nerves

Palpation: Deep muscles; cannot be readily palpated.

Synergists are the **multifidus, rotatores, semispinalis,** and **erector spinae** groups; their **antagonist** is the **rectus abdominis** muscle.

Intertransversarii
(in•ter•trans•ver•**sar**•e•eye)

POSTEROLATERAL VIEW

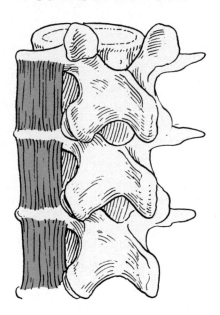

POSTERIOR VIEW

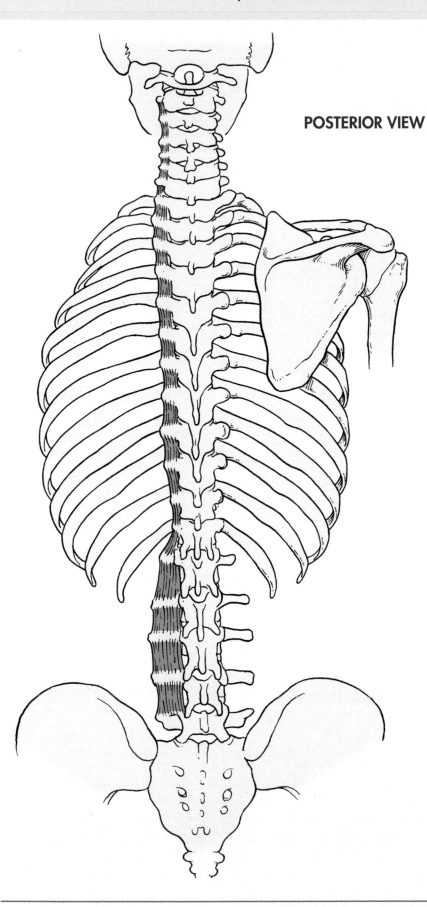

Origin (inferior attachment):
Transverse processes
of all vertebrae from
lumbar to axis

Insertion (superior attachment):
Transverse process
of next superior
vertebrae

Action: Lateral flexion of
vertebral column

Innervation: Ventral and dorsal
rami of spinal nerves

Palpation: Deep muscles; cannot
be readily palpated

The attachments to C1 and C2
are often absent. **Synergists** are
the **interspinales, rotatores,** and
multifidus groups.

Quadratus Lumborum
(kwa•drah•tus) (lum•bor•um)

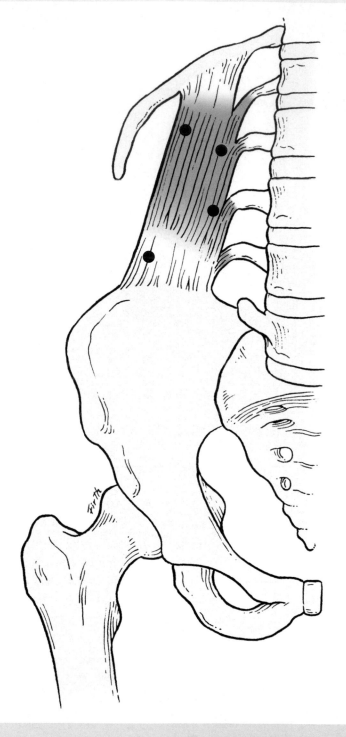

ANTERIOR VIEW

Origin (inferior attachment): Iliolumbar ligament and the posterior portion of the iliac crest

Insertion (superior attachment): Inferior border of last rib and the transverse processes of the first four lumbar vertebrae (L1–L4)

Action: Flexes lumbar region of vertebral column laterally to the same side. Both muscles together stabilize and extend the lumbar vertebrae and assist forced expiration.

Innervation: Ventral rami of the twelfth thoracic (T12) and upper three lumbar spinal nerves (L1–L3)

Palpation: In supine position, palpate deep in lumbar region above iliac crest during active elevation of hip.

The superior attachment may extend to the eleventh rib and also to L5. The **trigger points** are found laterally near the rib or iliac attachment. Its **referred pain pattern** is found in the gluteal and groin area and also in the sacroiliac joint and the greater trochanter. If the trigger points are active, a cough or sneeze can cause severe pain in the lower back. **Synergists** are the **erector spinae group, serratus posterior inferior, subcostalis, internal intercostals** and the **transversus thoracis**; antagonists are the **scalenes group, pectoralis minor, serratus posterior superior** and the **levatores costarum**.

Levatores Costarum

(lev•ah•**tor**•eez) (cos•**tar**•um)

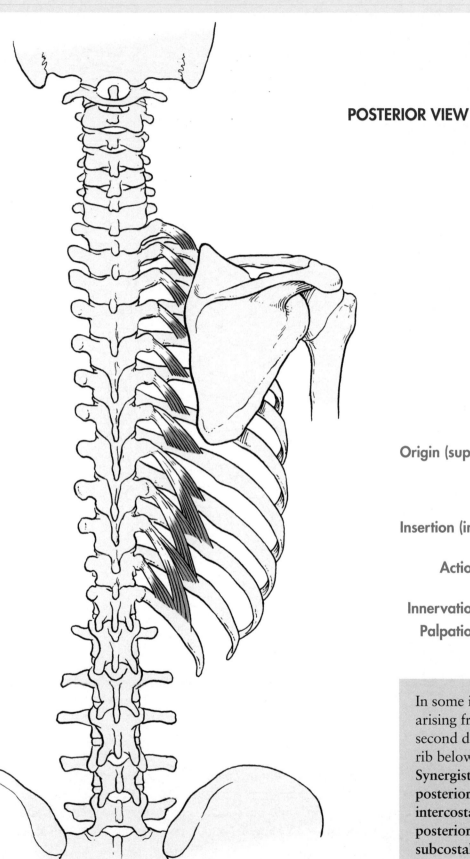

POSTERIOR VIEW

Origin (superior attachment): Transverse processes of seventh cervical and upper eleven thoracic vertebrae (C7–T11)

Insertion (inferior attachment): Laterally to costal angle of next lower rib

Action: Raises rib and laterally flexes and rotates vertebral column

Innervation: Intercostal nerves (T1–T11)

Palpation: Deep muscles; cannot be readily palpated.

In some individuals, the muscle slips arising from T8 to T10 may each have a second division that extends to the second rib below as shown on the illustration. **Synergists** are the **scalenes group**, **serratus posterior superior**, **subclavius** and **external intercostals**; **antagonists** are the **serratus posterior inferior**, **quadratus lumborum**, **subcostales**, and the **transversus thoracis.**

Intercostales Externi—External Intercostals

(in•ter•**cos**•tal•eez)(ex•**ter**•nye)

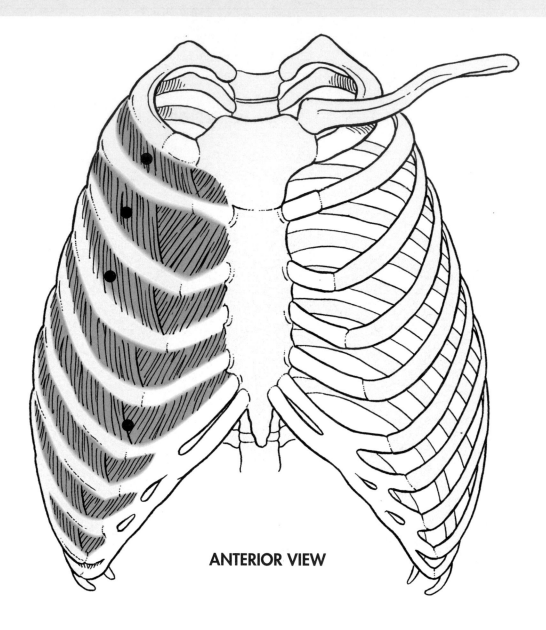

ANTERIOR VIEW

Origin (superior attachment): Lower margin of upper eleven ribs

Insertion (inferior attachment): Superior border of rib below

Action: With first ribs fixed by scalenes, they pull the ribs toward one another to elevate the rib cage.

Innervation: Intercostal nerves (T1–T11)

Palpation: Palpate between ribs.

The **external intercostal** muscles act synergistically with the diaphragm to aid in inspiration. The fibers are oriented obliquely down and forward toward the costal cartilages. In the lower intercostal spaces, the fibers are continuous with the **external oblique** muscle of the abdominal wall. The **trigger points** are located anteriorly between the ribs. **Synergists** are the **serratus posterior superior** and **scaleni** group; **antagonists** are the **internal intercostals**.

Intercostales Interni—Internal Intercostals

(in•ter•**cos**•tal•eez)(in•**ter**•nye)

ANTERIOR VIEW

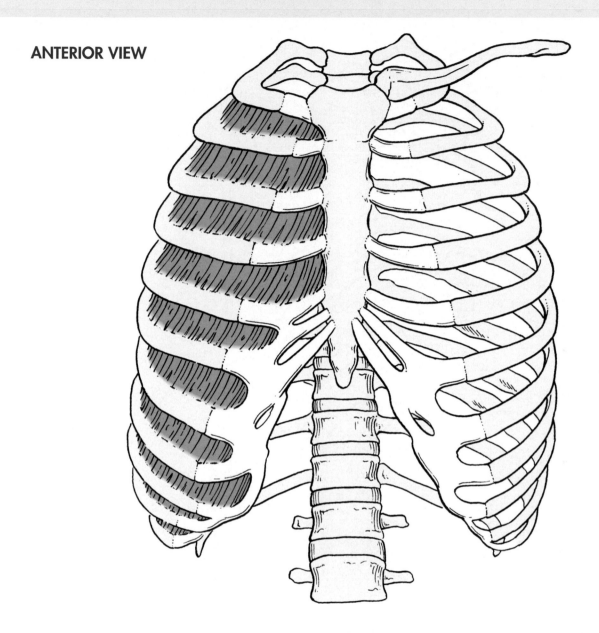

Origin (superior attachment): Ridge of inner surface of rib and corresponding costal cartilage

Insertion (inferior attachment): Superior border of rib below

Action: Draw ribs together and depress the rib cage

Innervation: Intercostal nerves (T1–T11)

Palpation: Palpate between costal cartilages.

The muscle fibers here are angled obliquely away from the costal cartilages. The contraction of these muscles decreases the size of the thoracic cavity and aids in forced expiration. **Synergists** are the **serratus posterior inferior** and **quadratus lumborum;** **antagonists** are the **external intercostals** and **scalenes.**

Subcostales—Subcostals
(sub•**cos**•tal•eez)

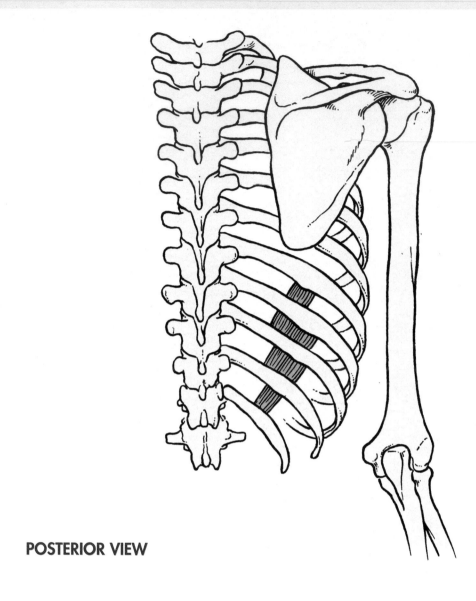

POSTERIOR VIEW

Origin (superior attachment): Inner surface of each rib near its angle

Insertion (inferior attachment): Medially on the inner surface of the second or third rib below

Action: Draws ventral part of ribs downward

Innervation: Intercostal nerve (T12)

Palpation: Deep muscles; cannot be palpated.

These muscles are deep to the internal intercostals. They are **synergistic** with the internal intercostals in decreasing the size of the thoracic cavity and aiding in forced expiration.

Transversus Thoracis

(trans•**ver**•sus) (tho•**ra**•sis)

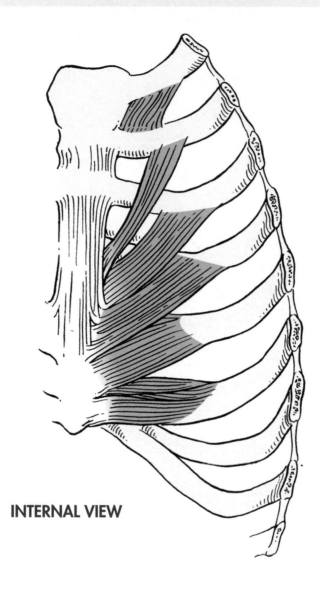

INTERNAL VIEW

Origin:	Inner surface of the body of the sternum, xiphoid process, and sternal ends of the costal cartilages of the last three or four true ribs
Insertion:	Inner surfaces of the costal cartilages of the second through sixth ribs
Action:	Draws ventral part of rib downward
Innervation:	Intercostal nerves (T1–T12)
Palpation:	Deep muscles; cannot be palpated.

This muscle, found on the inside of the rib cage, decreases the size of the **thoracic cavity** and thus aids in forced expiration. **Synergists** are the **quadratus lumborum, serratus posterior inferior, subcostales** and the **internal intercostals**; antagonists are the **scalenes group, serratus posterior superior, levatores costarum, subclavius** and the **external intercostals**.

Serratus Posterior Superior
(ser•**ray**•tus)

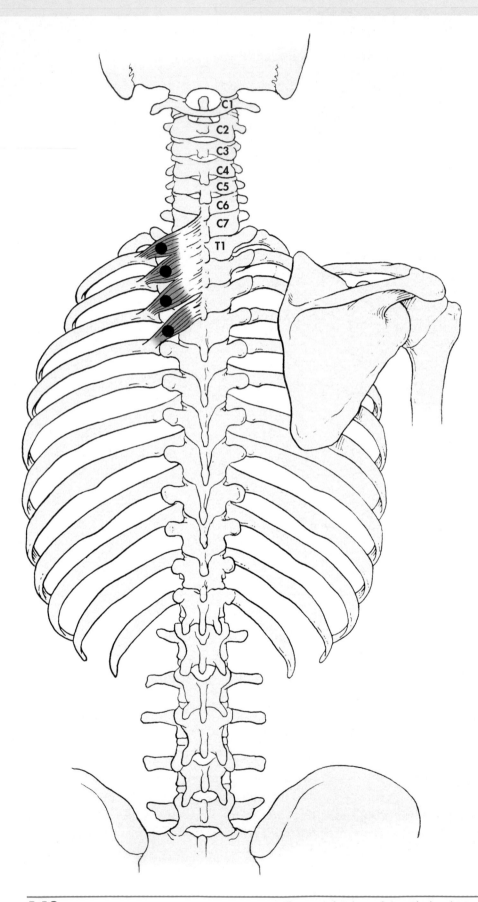

POSTERIOR VIEW

C1
C2
C3
C4
C5
C6
C7
T1

Origin: Lower portion of ligamentum nuchae and the spinous processes of the sixth and seventh cervical through the third thoracic vertebrae (C6–T3)

Insertion: Upper border and external surfaces of ribs two through five lateral to their angles

Action: Assists in raising ribs during inspiration

Innervation: Second through fifth intercostal nerves (T2–T5)

Palpation: Place palpating hand in region of upper rhomboids while taking a deep breath; difficult to palpate.

This muscle lies under the **rhomboideus** next to the ribs. The **trigger points** are under the scapula near the insertion of the muscle on the ribs. Its **referred pain pattern** is under the upper portion of the scapula. **Synergists** are the **scalenes group, levatores costarum, subclavius** and the **external intercostals**; antagonists are the **quadratus lumborum, serratus posterior inferior, subcostales, internal intercostals** and the **transversus thoracis.**

Serratus Posterior Inferior

(ser•**ray**•tus)

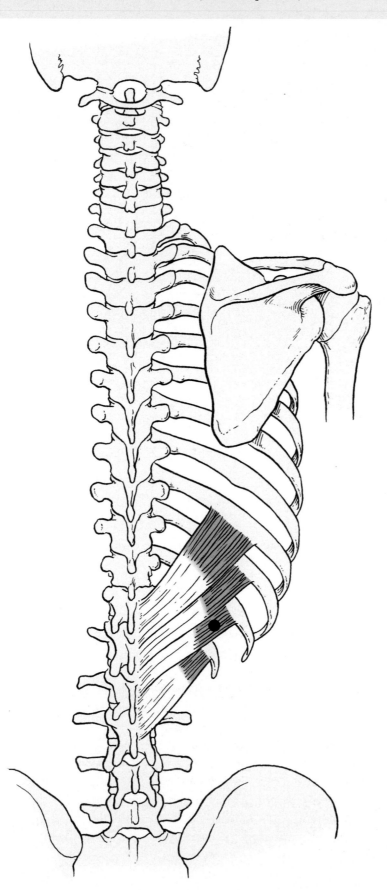

POSTERIOR VIEW

Origin:	Spinous process of the last two thoracic and upper three lumbar vertebrae (T11–L3)
Insertion:	Inferior borders and outer surfaces of lower four ribs just lateral to the angles
Action:	Depresses last four ribs (this is somewhat controversial in light of recent studies since it shows no electro-myographic activity during respiration)
Innervation:	The ninth through twelfth thoracic nerves (T7–T12)
Palpation:	Deep muscle; cannot be readily palpated.

The **trigger point** is in the belly of the muscle near the eleventh rib. Its **referred pain pattern** is a nagging ache in the area of the muscle. **Synergists** are the **internal intercostals, subcostales, transversus thoracis,** and the **quadratus lumborum; antagonists** are the **scalenes group, levatores costarum, subclavius, external intercostals,** and the **serratus posterior superior.**

Diaphragm
(dye•eh•fram)

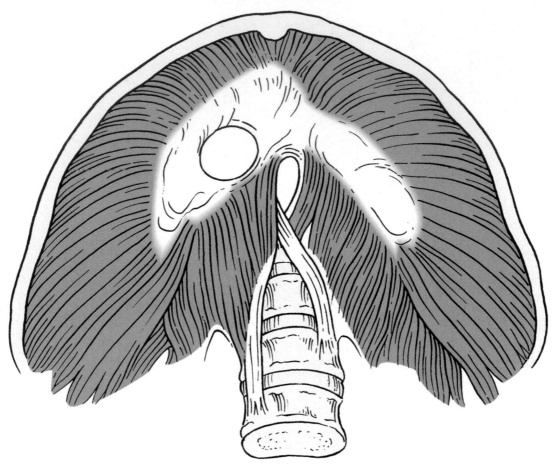

INFERIOR VIEW

Origin:	First three lumbar vertebrae, lower six costal cartilages, and inner surface of xiphoid process of sternum
Insertion:	Muscle fibers converge upward and inward to form the central tendon
Action:	Flattens on contraction, increasing the vertical dimensions of thorax
Innervation:	Phrenic nerve (C3–C5)
Palpation:	Palpate on inferior margin of anterior rib cage while taking a deep breath and slowly exhaling.

The **diaphragm** is the most important muscle of inspiration. When the muscle is relaxed, it is dome-shaped. It flattens as it contracts, increasing the volume of the thoracic cavity. The alternate contraction and relaxation causes pressure changes in the abdominopelvic cavity that assist in the return of venous blood and lymph to the heart. A protrusion of a structure through a weakened muscular layer is called a **hernia**. A **hiatal hernia** occurs because there is higher pressure in the abdomen than in the thorax, forcing part of the fundus of the stomach upwards through the esophageal hiatus into the thorax due to the weakening of the cardiac sphincter. Other examples of hernias are umbilical, femoral, and inguinal. The **Heimlich Maneuver** causes pressure on the diaphragm to increase intrathoracic pressure that forces food out of the laryngeal opening of a choking victim.

Rectus Abdominis
(**rek**•tus) (ab•**dom**•ih•nis)

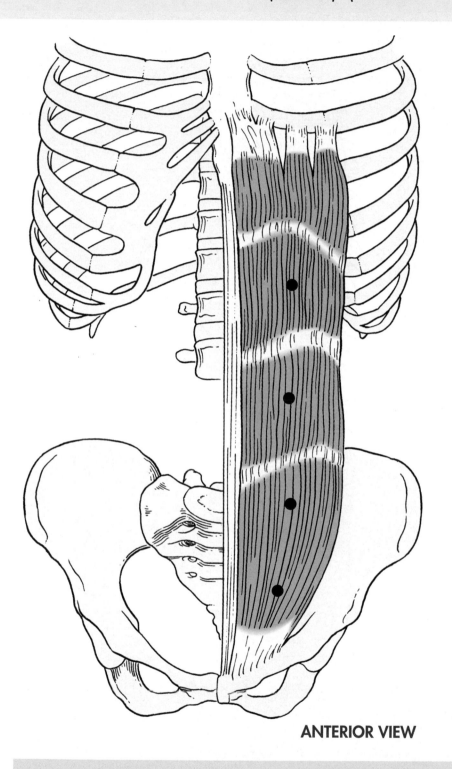

ANTERIOR VIEW

Origin (inferior attachment): Crest of pubis and pubic symphysis

Insertion (superior attachment): Cartilage of fifth, sixth, and seventh rib and xiphoid process of sternum

Action: Compresses the abdominal cavity and flexes the vertebral column

Innervation: Anterior primary rami of the seventh through twelfth intercostal nerves (T7–T12)

Palpation: Palpate on the anterior medial surface of abdomen during active flexion of trunk.

These are the "abs." Tendinous bands divide each **rectus** into four bellies. Each muscle is enclosed in a sheath formed from the aponeurosis of the lateral abdominal muscles. This muscle contracts strongly during "sit-ups" or when a person is lying in a supine position and raises the legs several inches from the floor. The **trigger points** are located in each belly near the linea alba. **Synergists** are the **external** and **internal obliques** and the **pyramidalis; antagonistic** are the **erector spinae muscles.**

Obliquus Externus Abdominis—External Oblique

(o•**bleek**•us) (ex•**tern**•us) (ab•**dom**•ih•nis)

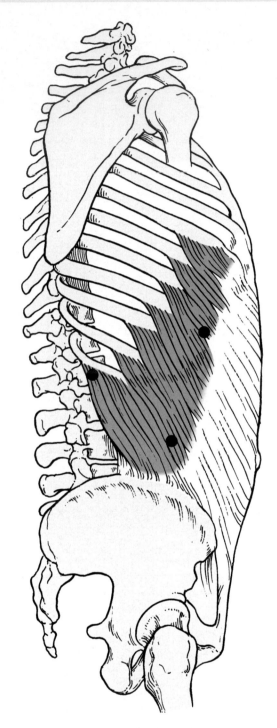

LATERAL VIEW

Origin: External surface of the lower eight ribs

Insertion: Anterior part of iliac crest and by abdominal aponeurosis to linea alba

Action: Compresses the abdominal cavity and laterally flexes and rotates vertebral column; both sides flex vertebral column anteriorly

Innervation: Ventral rami of the lower six thoracic nerves (T7–T12)

Palpation: Palpate on lateral side of abdomen during active trunk rotation.

This is the most superficial of the three side abdominal muscles. Its fibers angle obliquely downward and medially. **Synergists** are the **rectus abdominis, internal oblique,** and **psoas major; antagonists** are the **erector spinae** muscles.

Obliquus Internus Abdominis—Internal Oblique
(o•**bleek**•us) (in•**tern**•us) (ab•**dom**•ih•nis)

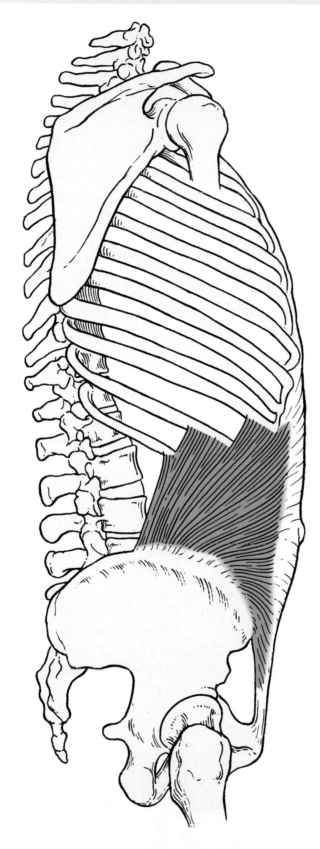

LATERAL VIEW

Origin: Lateral half of inguinal ligament, anterior two thirds of the iliac crest, and thoracolumbar fascia

Insertion: Upper fibers into cartilages of last three ribs, the remainder into the aponeurosis extending from the tenth costal cartilage to the pubic bone

Action: Compresses abdominal contents, laterally bends and rotates vertebral column; it also aids the rectus abdominis in flexing vertebral column

Innervation: Ventral rami of the lower six thoracic and first lumbar spinal nerves (T7–T12, L1)

Palpation: Deep muscle; cannot be readily palpated.

The **internal oblique** is important in forced expiration, coughing, and sneezing. Contraction squeezes the abdominal contents. It is the middle of the three layers of abdominal wall muscles. **Synergists** are the **external oblique** and **rectus abdominis; antagonists** are the **erector spinae** muscles.

Transversus Abdominis—Transverse Abdominal

(trans•**ver**•sus) (ab•**dom**•ih•nis)

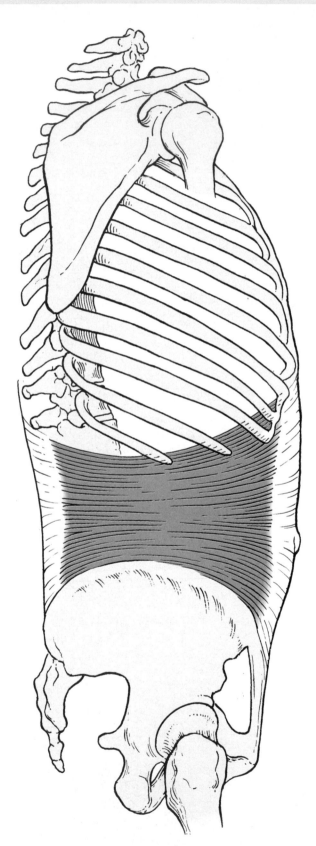

LATERAL VIEW

Origin: Lateral part of inguinal ligament, iliac crest, thoracolumbar fascia, and cartilage of lower six ribs

Insertion: Abdominal aponeurosis to linea alba

Action: Constricts the abdomen and supports the abdominal viscera

Innervation: Ventral rami of the lower six thoracic and first lumbar spinal nerves (T7–T12, L1)

Palpation: Deep muscle; cannot be readily palpated.

The **transversus abdominis** is the innermost of the three abdominal muscle layers. Its fibers run horizontally while the other two abdominal muscle layers' fibers run obliquely.

Cremaster

(kree•**mas**•ter)

ANTERIOR VIEW

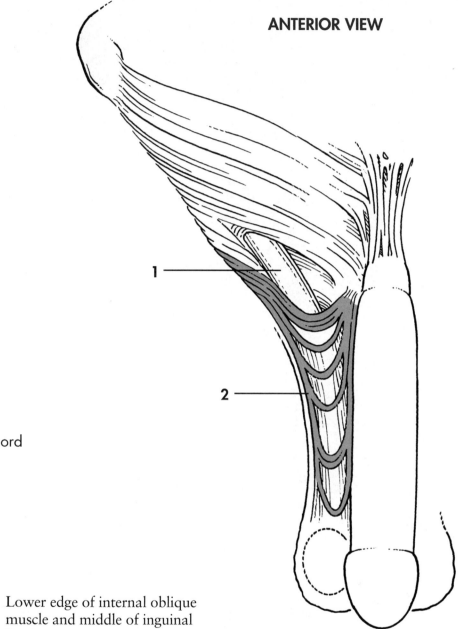

1 Spermatic cord
2 Cremaster

Origin:	Lower edge of internal oblique muscle and middle of inguinal ligament
Insertion:	Pubic tubercle and crest of pubis
Action:	Pulls the testes up toward the superficial inguinal ring
Innervation:	Genital branch of the genitofemoral nerve from the first and second lumbar nerves (L1, L2)

Contraction of the **cremaster muscle** reflexively elevates the testis to a higher position in the scrotum for warmth and to protect against injury. Under very warm conditions, the muscles relax, enabling the testis to sit lower with greater heat loss from the surrounding scrotal skin. These responses help maintain an optimal temperature in the testis for the production of male sex cells. Although a striated muscle, the cremaster is not usually under voluntary control.

Superficial Transverse Perineus

(soo•per•**fish**•al) (trans•**vers**) (perr•in•**nee**•us)

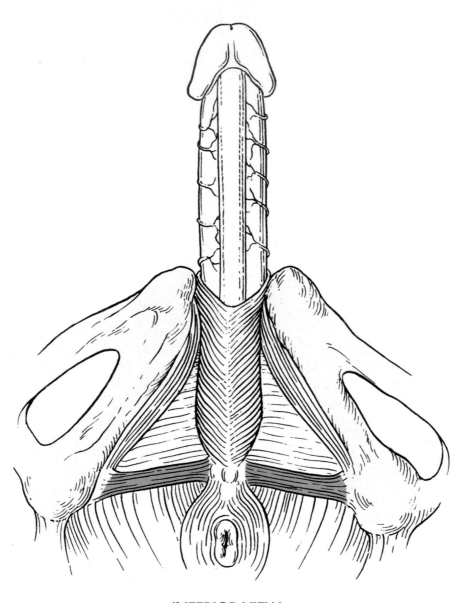

INFERIOR VIEW

Origin:	Medial and anterior part of the ischial tuberosity
Insertion:	Central tendinous point of perineum
Action:	Stabilizes and strengthens perineum
Innervation:	Perineal branches of pudendal nerve (S2–S4)

In males, these paired muscle bands are posterior to the urethral opening; in females, they are posterior to the vaginal opening. They are sometimes absent.

Coccygeus
(kok•sih•jee•us)

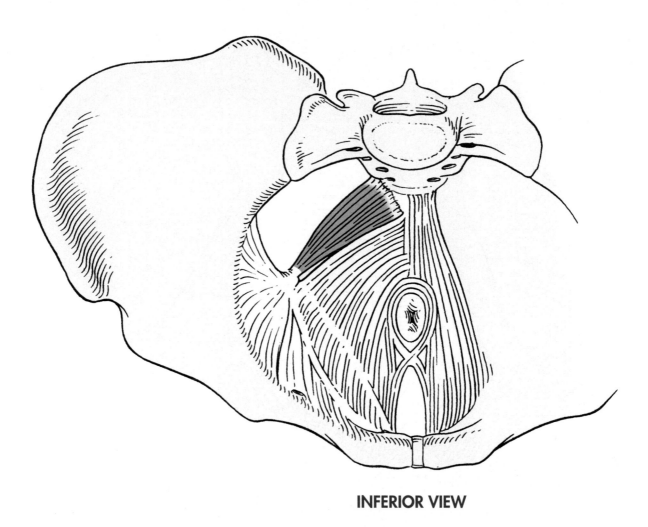

INFERIOR VIEW

Origin:	Pelvic surface of the ischial spine and the sacrospinous ligament
Insertion:	Margin of the coccyx and the lower sacrum
Action:	Supports pelvic viscera, supports coccyx and pulls it forward after it has been reflected by defecation or childbirth, and assists in closing the posterior part of the pelvic outlet
Innervation:	Fourth and fifth sacral nerves (S4, S5)

This is a small triangular muscle lying posterior to the **levator ani**. It forms the posterior part of the pelvic diaphragm.

Sphincter Urethrae
(sfink•ter) (yoo•ree•three)

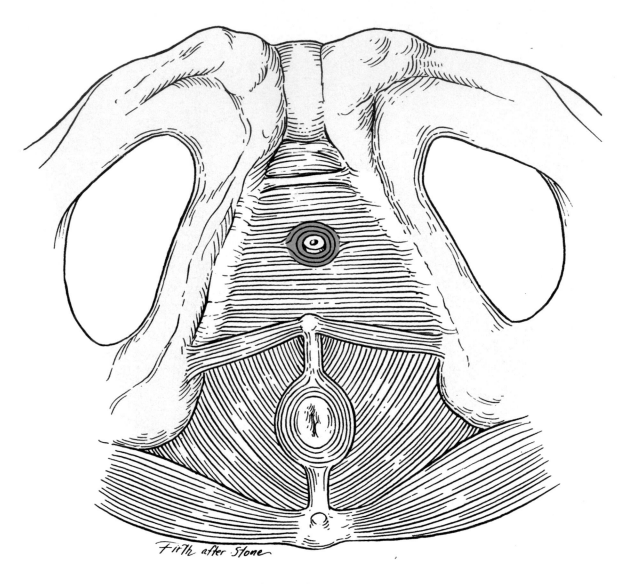

INFERIOR VIEW

Origin:	Ischiopubic rami
Insertion:	Midline raphe
Action:	Constricts urethra and helps support pelvic organs
Innervation:	Perineal branch of pudendal nerve (S2–S4)

In females, this muscle encircles the **urethra** and the **vagina** (not illustrated).

Levator Ani
(lev•**vay**•ter)(**ah**•nee)

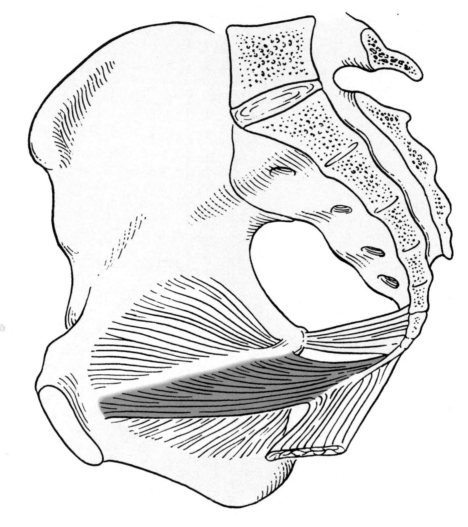

LATERAL—INTERNAL VIEW

Origin: Pelvic surfaces of the pubis, inner surface of the ischial spine and the obturator fascia

Insertion: Inner surface of the coccyx, levator ani of opposite side, and sides of rectum

Action: Forms the floor of the pelvic cavity, constricts the lower end of the rectum and vagina, and supports and slightly raises the pelvic floor

Innervation: Fourth sacral nerve and inferior rectal nerve (S4)

This is a broad, flat muscle whose fibers extend inferomedially, forming a muscular sling around the male **prostate** or female **vagina, urethra,** and **anorectal junction** before meeting in the median plane. It can be **palpated** internally in the female.

Ischiocavernosus

(is•key•o•kav•er•**no**•sus)

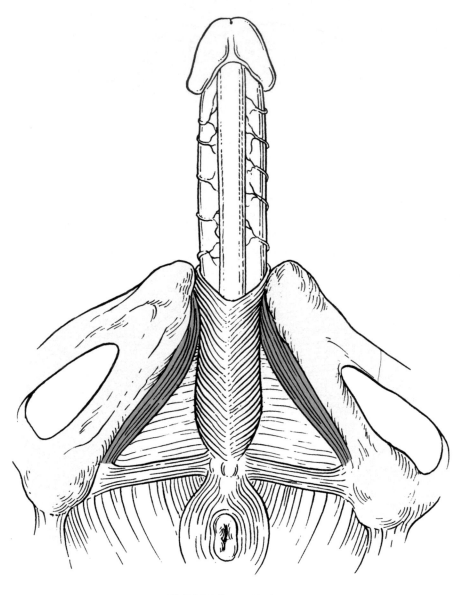

INFERIOR VIEW

Origin:	Inner surface of the ischial tuberosity and the ramus of the ischium
Insertion:	Aponeurosis on the sides and undersurface of the crus penis or clitoris
Action:	Compresses the crus penis, which obstructs venous return and maintains erection of penis or clitoris
Innervation:	Perineal branch of the pudendal nerve (S2–S4)

External Anal Sphincter
(ex•ter•nal) (ay•nal) (sfink•ter)

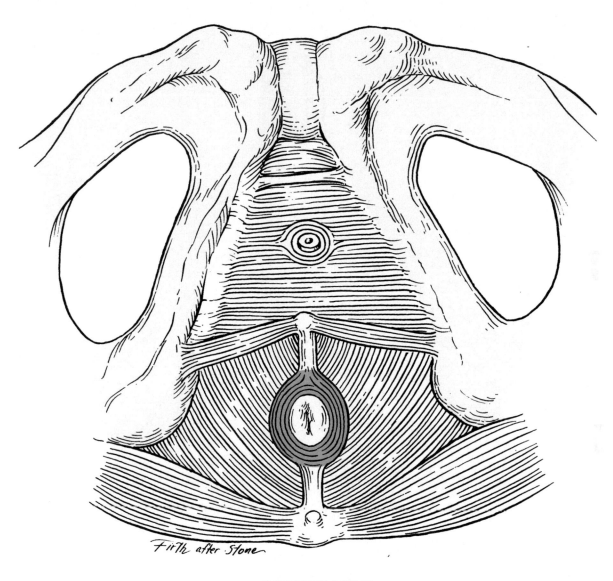

INFERIOR VIEW

Origin:	Central tendon of perineum
Insertion:	Midline raphe and the coccygeus muscle
Action:	Voluntary muscle circling the anus preventing defecation
Innervation:	Pudendal nerve (S2–S4)

Bulbospongiosus
(bul•bo•spon•jee•**oh**•sus)

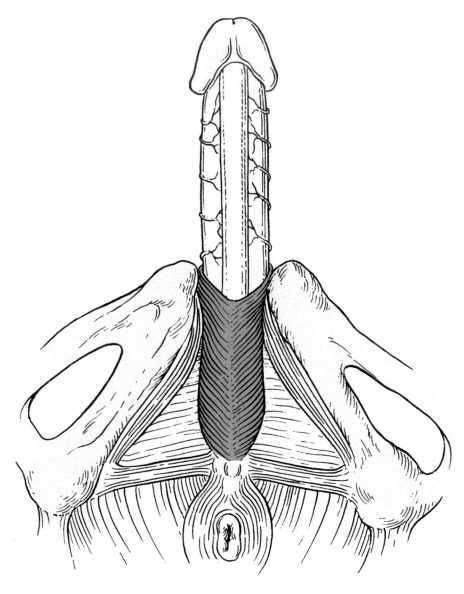

INFERIOR VIEW

Origin:	Central tendon of the perineum and midline raphe of male penis
Insertion:	Anteriorly into corpus cavernosa of penis or clitoris
Action:	Empties male urethra and assists in erection of penis in male and clitoris in female
Innervation:	Perineal branch of pudendal nerve (S2–S4)

Muscles of the Shoulder and Upper Arm

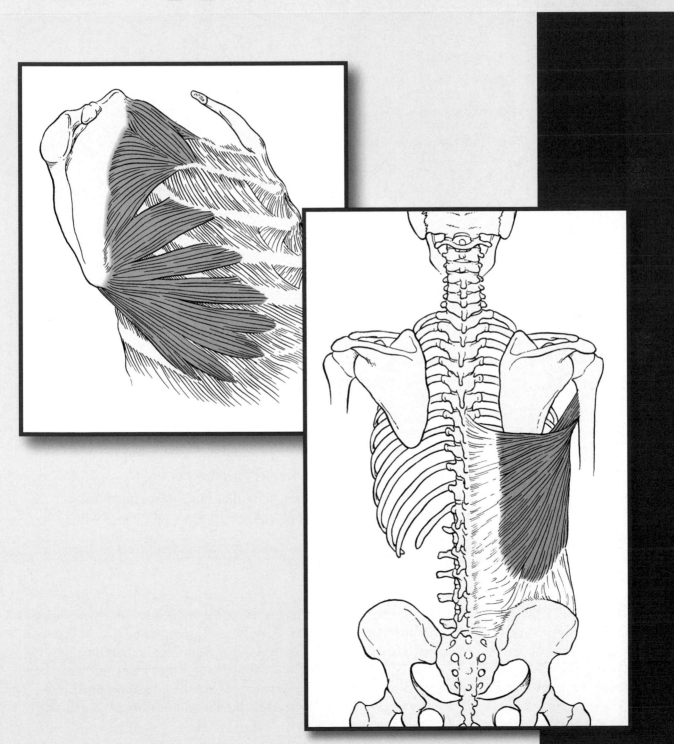

7

Pectoralis Major
(pek•toh•ral•lis)

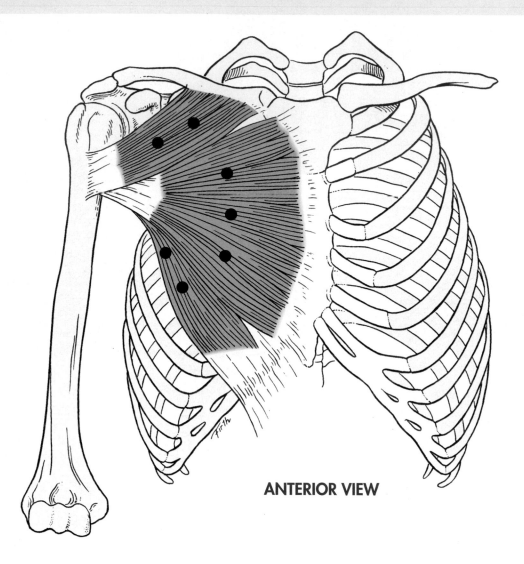

ANTERIOR VIEW

Origin (proximal attachment): Ventral surface of the sternum down to the seventh rib, sternal half of clavicle, cartilage of true ribs, and aponeurosis of the external oblique muscle

Insertion (distal attachment): Lateral lip of the intertubercular groove of the humerus

Action: Protracts scapula, adducts and medially rotates arm, clavicular head flexes humerus, sternal head extends humerus, and with insertion fixed it assists in elevation of the thorax

Innervation: Medial and lateral pectoral nerves (C5–C8, T1))

Palpation: Palpate along axillary border during active adduction of arm.

Poland's Syndrome is typically characterized by absence of pectoral muscles on the right side combined with abnormalities in the skeletal structure and skin, fusion of the digits (syndactyly) and assorted shoulder/thoracic bone anomalies. The **trigger points** for this muscle are in the belly for each portion, and the **referred pain pattern** is the chest and breast down to the ulnar aspect of the arm to the fourth and fifth fingers. In "pull-ups," it pulls the thorax up to the fixed arm position. Sternal division **synergists** are the **latissimus dorsi, sub-scapularis,** and **teres major;** clavicular division synergists are the **biceps brachii, latissimus dorsi,** and **deltoid.**

Pectoralis Minor
(pek•toh•ral•lis)

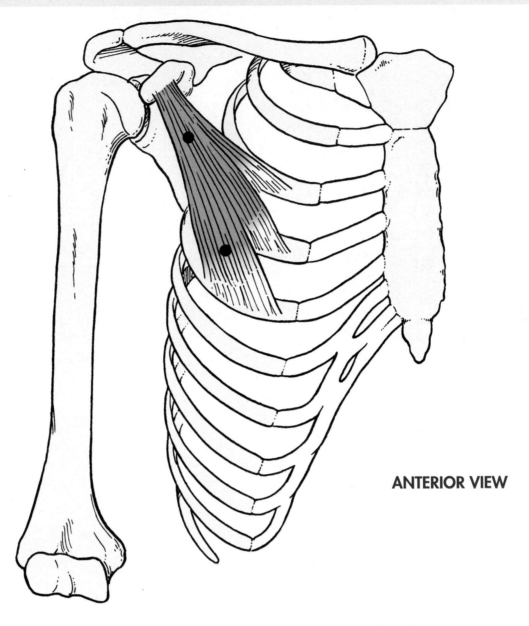

ANTERIOR VIEW

Origin (proximal attachment): Anterior surfaces of the third through fifth rib

Insertion (distal attachment): Coracoid process of the scapula

Action: With ribs fixed, it draws the scapula forward and downward, and with scapula fixed, it draws the rib cage superiorly

Innervation: Medial pectoral nerve (C8, T1)

Palpation: With arm behind and lifted away from back causing extension at shoulder, palpate just inferior to coracoid process.

The **trigger points** for this muscle are near the insertion at the ribs and at the coracoid process. Its **referred pain pattern** is the front of the chest and down the ulnar side of the arm and mimics the symptoms of angina. **Synergists** are the **pectoralis major** and **serratus anterior**.

Sternalis
(ster•nal•is)

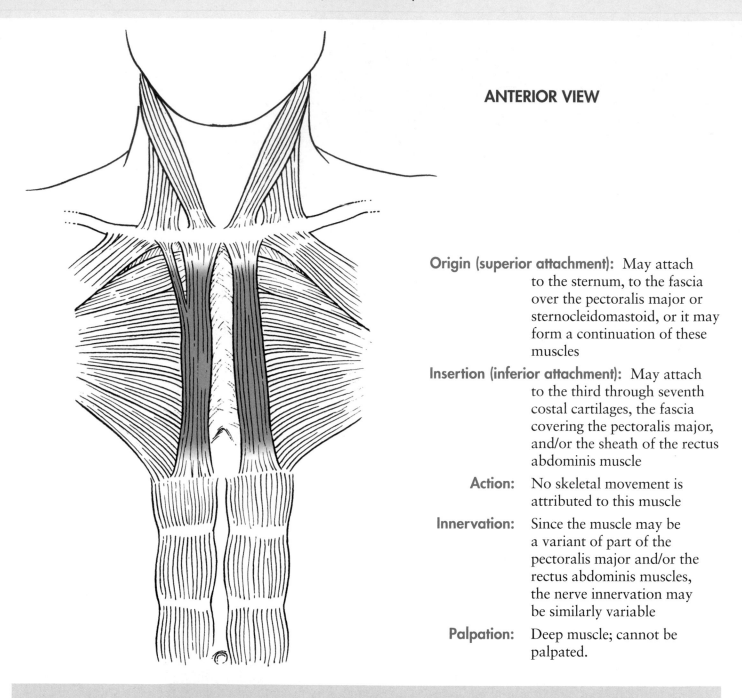

ANTERIOR VIEW

Origin (superior attachment): May attach to the sternum, to the fascia over the pectoralis major or sternocleidomastoid, or it may form a continuation of these muscles

Insertion (inferior attachment): May attach to the third through seventh costal cartilages, the fascia covering the pectoralis major, and/or the sheath of the rectus abdominis muscle

Action: No skeletal movement is attributed to this muscle

Innervation: Since the muscle may be a variant of part of the pectoralis major and/or the rectus abdominis muscles, the nerve innervation may be similarly variable

Palpation: Deep muscle; cannot be palpated.

The **sternalis** occurs in 1 in 20 people. It is a band of fibers of varying length, width, and thickness that may occur unilaterally and bilaterally (see illustration) or be fused medially along the sternum. **Trigger points,** more common on the left side, may be positioned anywhere along its length, from the manubrium to the xiphoid process, and on either or both sides or in the midline of the sternum when the muscle fuses across the sternum. The **referred pain pattern** includes the entire sternal and substernal region and may extend on the same side across the upper pectoral area and front of the shoulder to the underarm and to the ulnar aspect of the elbow. The pattern is similar to the substernal ache of **myocardial infarction** or **angina pectoris.** Clinically, the **sternalis** has the potential to initially cause misdiagnosis of breast tumors or their recurrence post-treatment.

Subclavius
(sub•**klay**•vee•us)

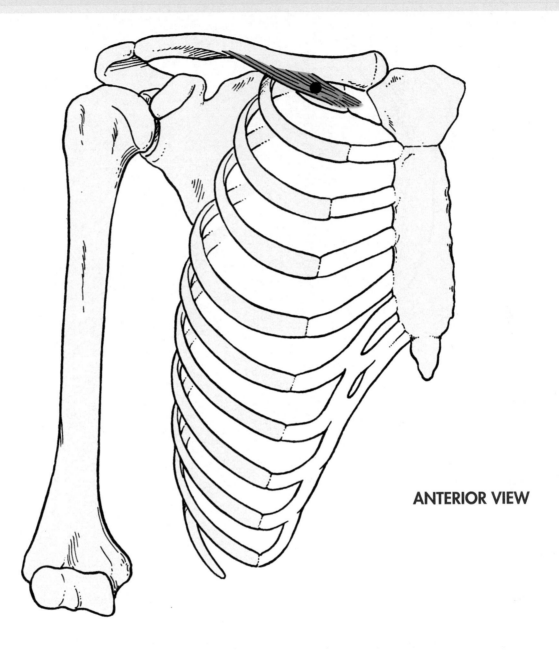

ANTERIOR VIEW

Origin (proximal attachment): Junction of the first rib and its costal cartilage

Insertion (distal attachment): Groove on the inferior surface of clavicle

Action: Depresses clavicle and draws shoulder forward and downward

Innervation: C5–C6

Palpation: Deep muscle; difficult to distinguish from Pectoralis Major.

The **trigger point** is in the belly of the muscle. **Synergists** are the **scalenes group, serratus posterior superior** and the **levatores costarum; antagonists** are the **superior trapezius** and the **supraspinatus.**

Serratus Anterior
(serr•**ray**•tus)

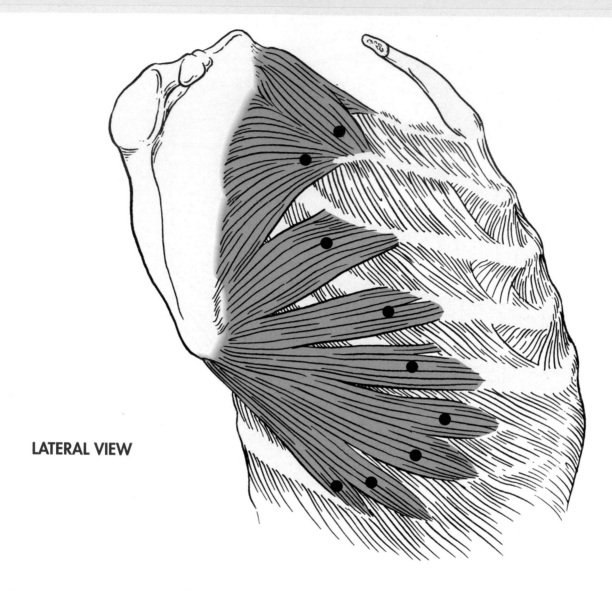

LATERAL VIEW

Origin (proximal attachment): Outer surfaces and superior borders of the first eight or nine ribs

Insertion (distal attachment): Anterior surface of the medial border of the scapula

Action: Abducts, protracts, and upwardly rotates the scapula; it also holds the scapula firmly against the thorax

Innervation: Long thoracic nerve (C5–C7)

Palpation: Palpate along anterior surface of ribs below the axilla during active scapula protraction.

The number of slips may vary between seven and twelve. The most superior slip may not attach to the first rib, and muscle slips may be continuous with other adjacent muscles. This muscle is important in horizontal arm movement such as pushing and punching. It is sometimes called the "boxer's muscle." Its **trigger points** are along the midaxillary line near the ribs. Its **referred pain pattern** is along the side and back of chest and down the ulnar aspect of the arm into the hand. Its **synergist** is the **pectoralis minor; antagonists** are the **lower trapezius, latissimus dorsi, rhomboids,** and the **pectoralis major.**

Trapezius
(tra•**pee**•zee•us)

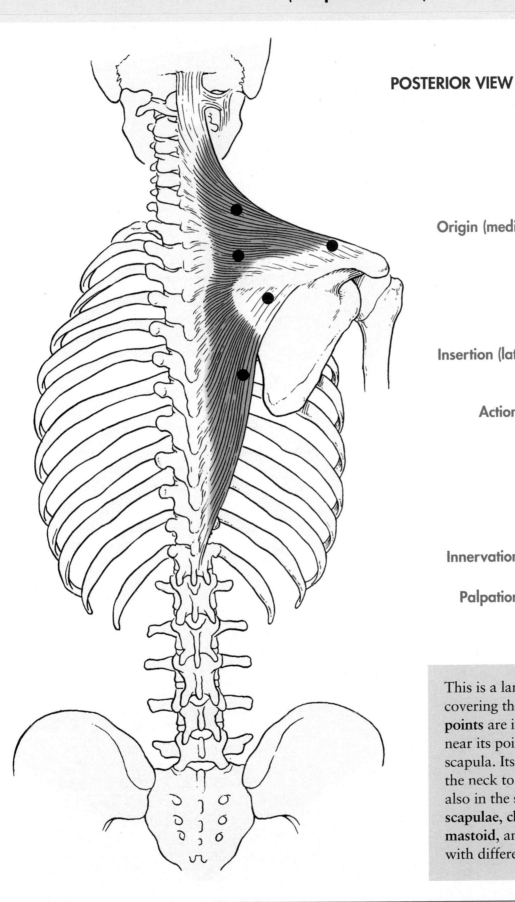

POSTERIOR VIEW

Origin (medial attachment): Medial third of superior nuchal line, external occipital protuber-ance, ligamentum nuchae, spinous process of the seventh cervical, and all thoracic vertebrae

Insertion (lateral attachment): Lateral third of clavicle, acromion process, and spine of scapula

Action: Stabilizes, raises, retracts, and rotates the scapula. The superior fibers elevate, the middle fibers retract, and the inferior fibers depress the scapula. Together the upper and lower fibers rotate the arm.

Innervation: Spinoaccessory (XI) and ventral rami of C3, C4

Palpation: Palpate its different regions during elevation, retraction, and depression of scapula.

This is a large triangular-shaped muscle covering the upper back. Its **trigger points** are in the belly of the muscle and near its points of attachment to the scapula. Its **referred pain patterns** are in the neck to the ear and the temple and also in the subscapular area. The **levator scapulae**, **clavicular head of sternocleido-mastoid**, and **rhomboids** are **synergistic** with different portions of the trapezius.

Levator Scapulae
(lev•**vay**•ter)(**skap**•yoo•lee)

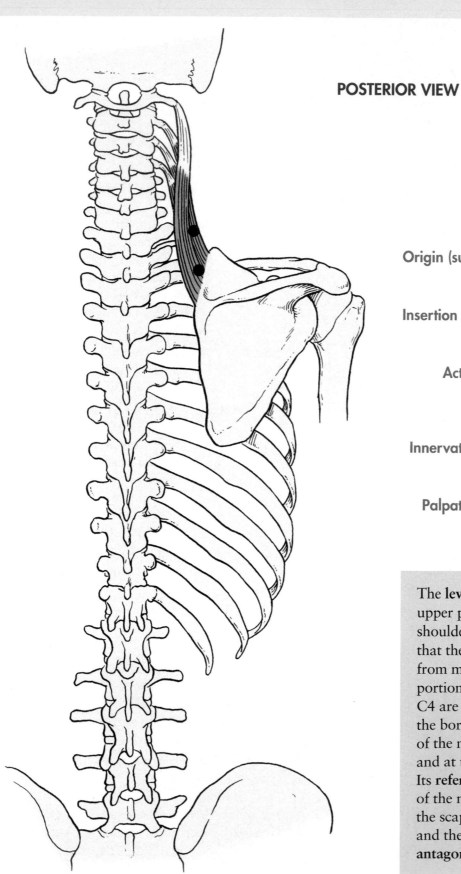

POSTERIOR VIEW

Origin (superior/medial attachment): Transverse processes of the first four cervical vertebrae (C1–C4)

Insertion (inferior/lateral attachment): Vertebral border of the scapula between the superior angle and the spine

Action: Raises scapula and draws it medially; with the scapula fixed, bends the neck laterally and rotates it to the same side

Innervation: Third and fourth cervical spinal nerves and dorsal scapular nerve (C3–C5)

Palpation: Palpate neck between sterno-cleidomastoid and trapezius during elevation of scapula.

The **levator** is an elevator. It works with the upper portion of the **trapezius** when the shoulders are shrugged. It has a twist in it so that the attachments at the atlas and axis are from muscle fibers that attach to the inferior portion of the border and the attachment at C4 are from fibers at the superior portion of the border. Its **trigger points** are at the belly of the muscle just as it begins its rotation and at the attachment point of the scapula. Its **referred pain pattern** is along the angle of the neck and the vertebral border of the scapula. **Synergists** are the **rhomboids** and the **trapezius**; the **latissimus dorsi** acts **antagonistically** to depress the scapula.

Rhomboid(eus) Major
(rom•boid) (ee•us)

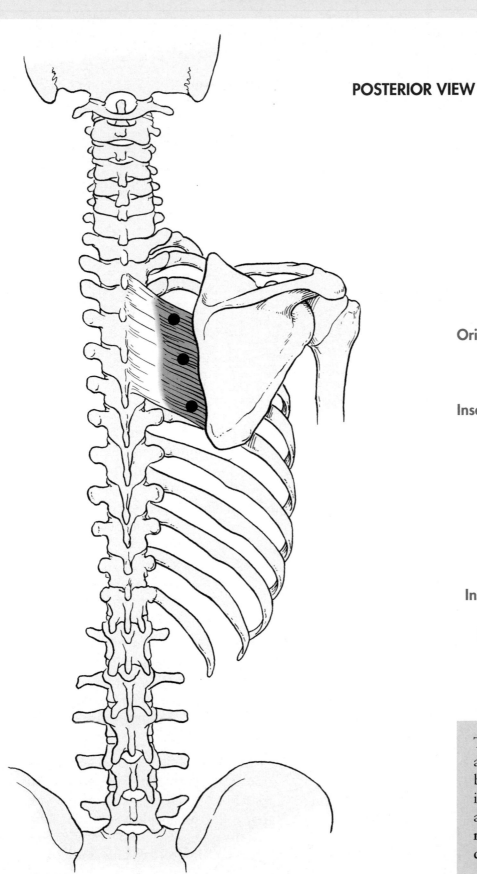

POSTERIOR VIEW

Origin (medial attachment): Spinous process of the second through fifth thoracic vertebrae (T2–T5)

Insertion (lateral attachment): Medial border of scapula between the spine and the inferior angle

Action: Adducts, retracts, elevates, and rotates the scapula so that the glenoid cavity faces downward and stabilizes the scapula

Innervation: Dorsal scapular nerve (C5)

Palpation: Palpate along vertebral border during active retraction and adduction of scapula.

The **trigger points** are found at the attachment point near the scapula border. Its **referred pain pattern** is in the scapula region. **Synergists** are the **trapezius, levator scapulae, rhomboid minor,** and **latissimus dorsi.**

Rhomboid(eus) Minor

(**rom**•boid) (ee•us)

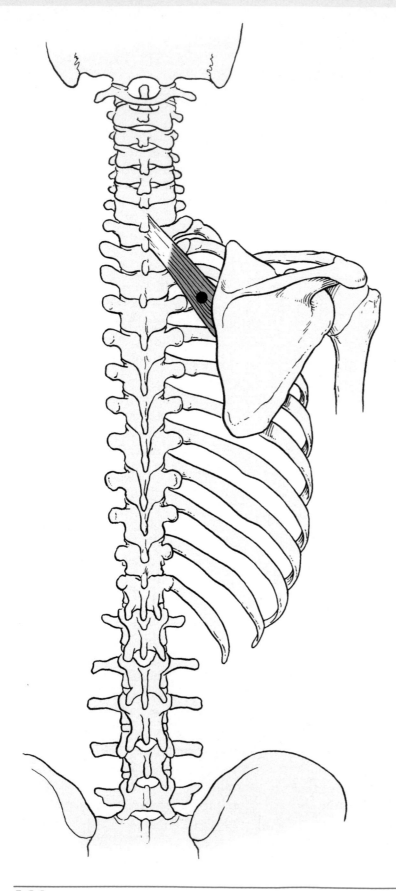

Origin (medial attachment): Spinous processes of the seventh cervical and first thoracic vertebrae (C7–T1)

Insertion (lateral attachment): Medial border of the scapula at the root of the spine

Action: Retracts and stabilizes the scapula, elevates the vertebral border, and rotates the scapula to depress the inferior angle

Innervation: Dorsal scapular nerve (C5)

Palpation: Palpate along vertebral border during active retraction and adduction of scapula.

The fibers of both **rhomboideus** muscles are arranged in an oblique, downward pattern. The **trigger points** are the same as for the **rhomboideus major**. Synergists are the **trapezius, levator scapulae, rhomboid major,** and **latissimus dorsi.**

Latissimus Dorsi
(la•**tiss**•eh•mus)(**dor**•sye)

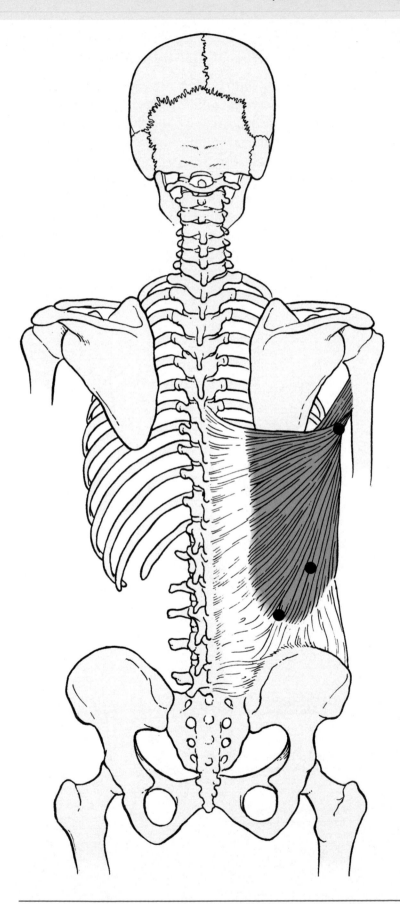

POSTERIOR VIEW

Origin: Indirect attachment through lumbodorsal fascia into spinous process of lower six thoracic and lumbar vertebrae (T7–L5), lower three to four ribs, and iliac crest

Insertion: In floor of intertubercular groove of humerus

Action: Extends, adducts, and medially rotates the arm; draws the shoulder downward and backward

Innervation: Thoracodorsal nerve (C6–C8)

Palpation: Palpate along lateral side of rib cage during active extension of the humerus.

This muscle may occasionally arise in part from the inferior angle of the scapula. This muscle is important in bringing the arm down in a power stroke as in hammering, swimming (crawl stroke), and rowing. Its **trigger points** are in the belly of the muscle near the rib attachments. Its **referred pain pattern** is below the scapula and into the ulnar side of the arm and the abdominal oblique area. It also acts as an accessory muscle of respiration. **Synergists** are the **rhomboids, pectoralis major**, and **teres major.**

The latissimus dorsi as a possible cardiovascular assist muscle.

Although there are effective mechanical heart assist devices today, in the mid 1980s two surgical techniques were developed for using the **latissimus dorsi** muscle to compensate for weak cardiac muscle contraction. The **latissimus** (Figure 1) was cut (Figure 2) and the anterior portion used either to wrap around the heart (Figure 3) or to form a pouch (lined by GORE-TEX®) into which the cut ends of the dorsal aorta (with artificial GORE-TEX® connections) were diverted (Figure 4). In coordination with the heart beat, a mechanical pacemaker stimulated contraction of the muscle wrapped around the ventricle or forming the pouch to increase the force with which the blood is propelled through the circulatory system. Advantages of both techniques included no risk of rejection or infection (the only "artificial" component was the GORE-TEX® and pacemaker, both compatible with body tissues), freedom from an external power source, and no time loss seeking heart donors.

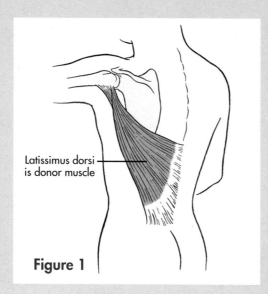

Latissimus dorsi is donor muscle

Figure 1

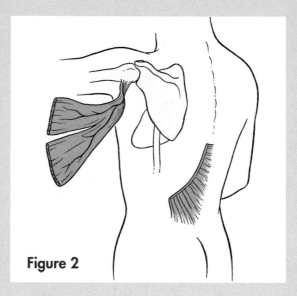

Figure 2

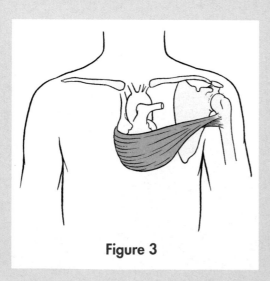

Figure 3

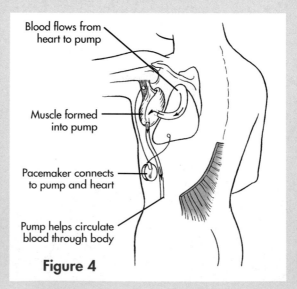

Blood flows from heart to pump

Muscle formed into pump

Pacemaker connects to pump and heart

Pump helps circulate blood through body

Figure 4

Deltoid(eus)
(del•toid) (ee•us)

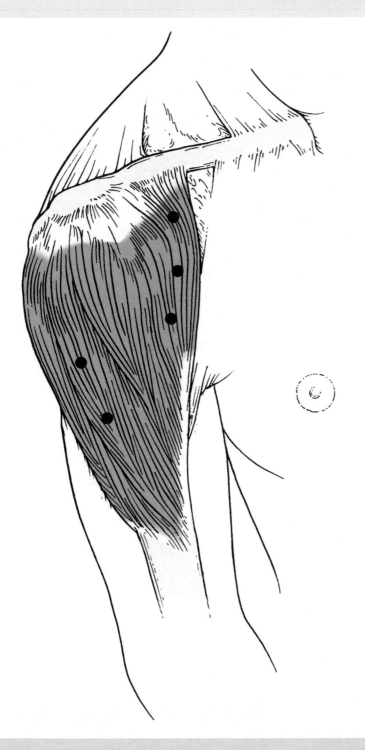

ANTEROLATERAL VIEW

Origin (proximal/medial): **Anterior portion**—superior surface of lateral third of clavicle

Middle portion—lateral border of acromion process of scapula

Posterior portion—lower border of the crest of the spine of the scapula

Insertion (distal/lateral): Deltoid tuberosity of the humerus

Action: **Anterior portion**—flexion and medial rotation of the arm

Middle portion—abducts the arm

Posterior portion—extends and laterally rotates the arm

Innervation: Axillary nerve (C5, C6)

Palpation: Palpate different heads during active adduction, abduction and extension of humerus.

The three portions may be separated from one another, and in some instances the clavicular and acromial portions may be absent. This shoulder muscle is one of the prime injection sites. It is active during the rhythmic arm swinging movements involved in walking. Its **trigger points** are found in the belly of the muscle. Its **referred pain pattern** is in the deltoid region and down the lateral surface of the arm. **Synergists** are the **supraspinatus** for **abduction** and **clavicular division of the pectoralis major** for **flexion**; **antagonists** are the **latissimus dorsi** and **sternal division of the pectoralis major** for **adduction**.

Subscapularis
(sub•skap•yoo•**lair**•is)

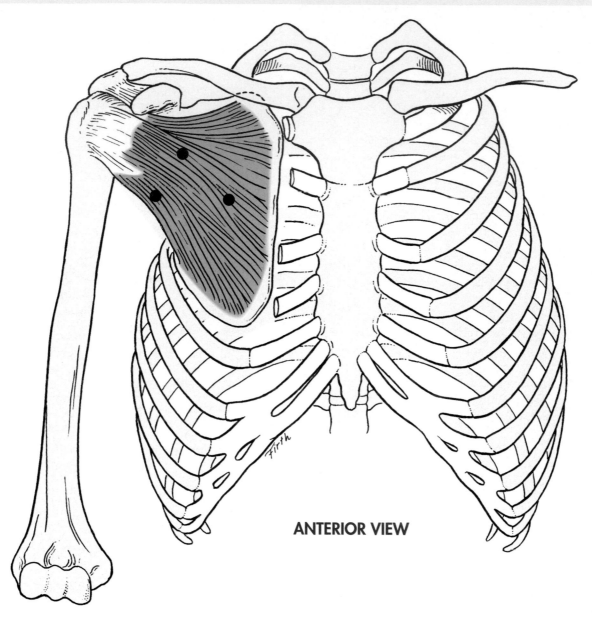

ANTERIOR VIEW

Origin (proximal/medial): Subscapular fossa of the scapula.

Insertion (distal/lateral): Lesser tubercle of the humerus and the ventral part of the capsule of the shoulder joint

Action: Medially rotates and stabilizes the head of the humerus in the glenoid cavity

Innervation: Upper and lower subscapular nerves (C5, C6)

Palpation: Cannot be readily palpated because of its position.

This "rotator cuff" muscle is often implicated in "frozen shoulder syndrome." Its **trigger points** are found near the attachment of the humerus and in its belly. Its **referred pain pattern** is in the posterior deltoid/triceps region down to the wrist. **Synergists** are the **teres major, pectoralis major,** and **latissimus dorsi; antagonists** are the **posterior deltoid, teres minor,** and the **infraspinatus.**

Supraspinatus
(su•prah•spy•**nay**•tus)

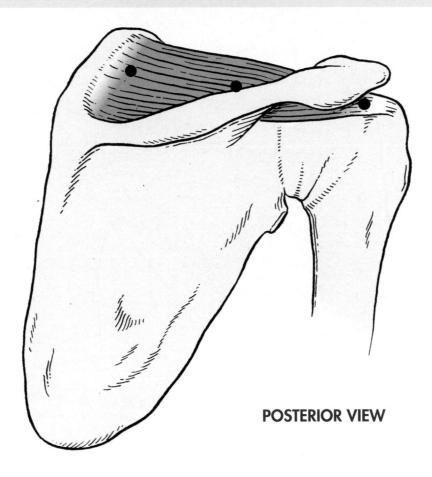

POSTERIOR VIEW

Origin (proximal/medial): Supraspinous fossa of the scapula

Insertion (distal/lateral): Superior part of the greater tubercle of the humerus and the capsule of the shoulder joint

Action: Abducts the arm and acts to stabilize the humeral head in the glenoid cavity during movements of the shoulder joint; one of four "rotator cuff" muscles

Innervation: Suprascapular nerve (C5, C6)

Palpation: Palpate above the spine of scapula during active abduction of arm.

This is one of the four "**rotator cuff**" muscles, and it is the tendon of this muscle that frequently is torn. Pain is described as deep in the shoulder and becomes progressively worse during abduction. The **trigger points** are near the tendons and in the belly of the muscle. The **referred pain pattern** is deep in the shoulder and down the arm to the elbow. **Synergists** are the **middle deltoid** and **infraspinatus**; antagonists are the **coraco-brachialis, sternal head of pectoralis major, latissimus dorsi,** and the **teres major.**

Infraspinatus
(in•frah•spy•**nay**•tus)

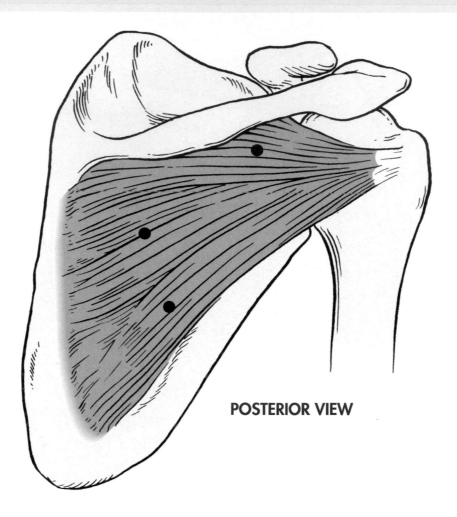

POSTERIOR VIEW

Origin (proximal/medial): Infraspinous fossa of the scapula

Insertion (distal/lateral): Middle part of the greater tubercle of the humerus and the capsule of the shoulder joint

Action: Lateral rotation of the shoulder and acts to stabilize the humeral head in the glenoid cavity; it abducts the humerus. It is one of the "rotator cuff" muscles

Innervation: Suprascapular nerve (C5, C6)

Palpation: Palpate below spine of scapula at axillary border during active lateral rotation of humerus.

This "rotator cuff" muscle has **trigger points** in the belly of the muscle below the spine of the scapula and near the medial border of the scapula. The **referred pain pattern** is deep into the shoulder and deltoid area down into the arm. **Synergists** are the **teres minor, subscapularis, supraspinatus** and the **middle and posterior deltoids; antagonists** are the **anterior deltoid, latissimus dorsi, teres major, pectoralis major,** and the **subscapularis.**

Teres Major
(ter•eez)

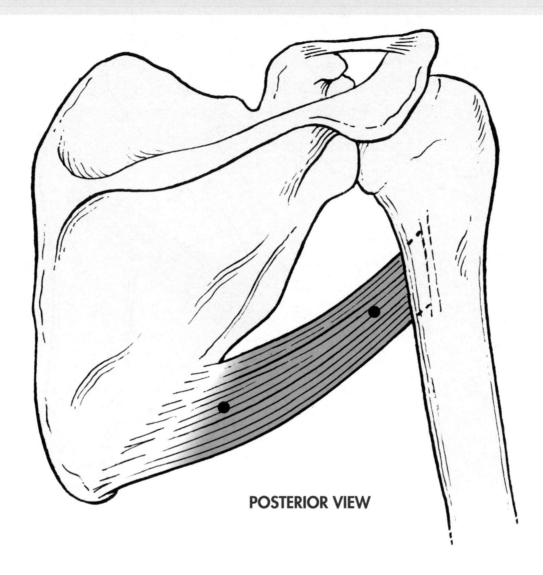

POSTERIOR VIEW

Origin (proximal/medial): Lower third of the posterior surface of the lateral border of the scapula

Insertion (distal/lateral): Medial lip of the bicipital (intertubercular) groove of the humerus

Action: Medially rotates, adducts, and extends the humerus

Innervation: Lower subscapular nerve (C6, C7)

Palpation: Palpate from inferior angle of scapula upward during active extension of humerus.

The **trigger points** are near both attachments. Its **referred pain pattern** is in the posterior deltoid region and down the dorsal surface of the arm. **Synergists** are the **anterior deltoid, pectoralis major, latissimus dorsi,** and the **subscapularis; antagonists** are the **posterior deltoid, teres minor, supraspinatus** and the **infraspinatus.**

Teres Minor
(ter•eez)

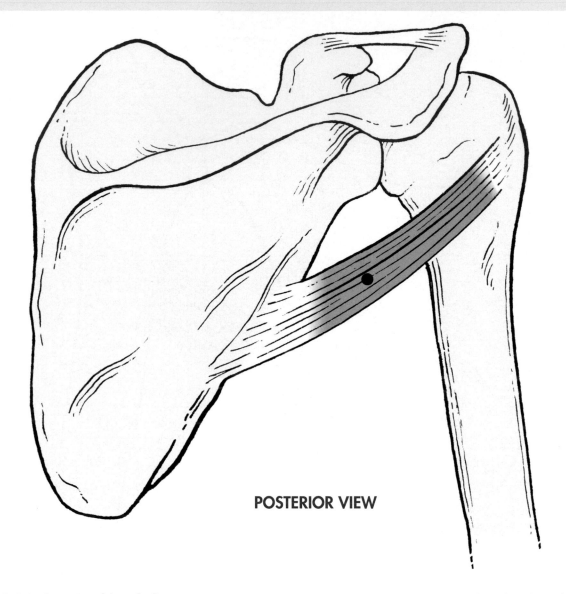

POSTERIOR VIEW

Origin (proximal/medial): Upper two thirds of the dorsal surface of the axillary border of the scapula

Insertion (distal/lateral): The capsule of the shoulder joint and the inferior part of the greater tubercle of the humerus

Action: Laterally rotates the arm and draws the humerus toward the glenoid cavity

Innervation: Axillary nerve (C5)

Palpation: Palpate along axillary border of scapula immediately below infraspinatus during active lateral rotation of humerus.

This is another one of the "rotator cuff" muscles. Its **trigger point** is in the belly of the muscle near its point of attachment. Its **referred pain pattern** is in the deltoid region. **Synergists** are the **posterior deltoid** and **infraspinatus**; antagonists are the **anterior deltoid, latissimus dorsi, teres major, pectoralis major,** and the **subscapularis.**

Coracobrachialis
(kor•ah•ko•bray•key•al•is)

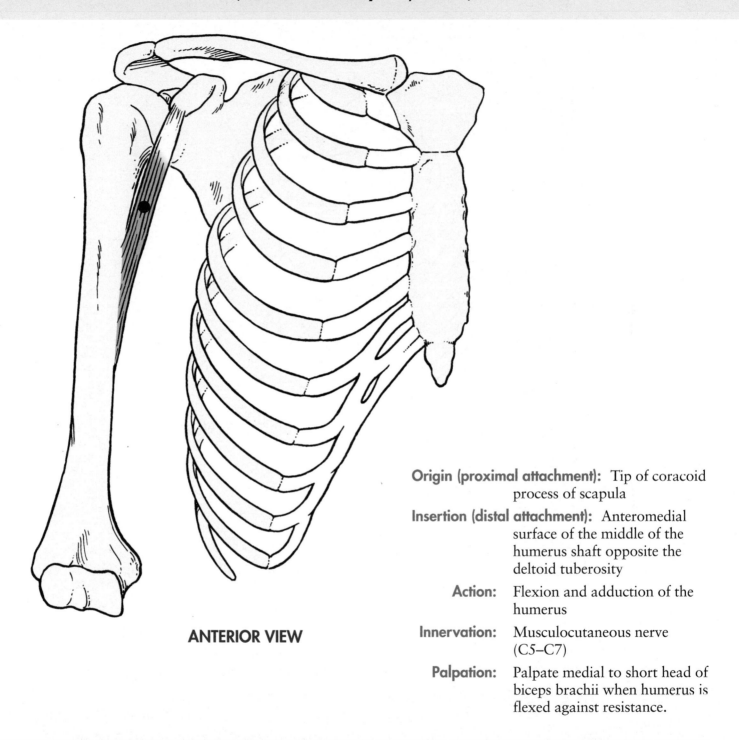

ANTERIOR VIEW

Origin (proximal attachment): Tip of coracoid process of scapula

Insertion (distal attachment): Anteromedial surface of the middle of the humerus shaft opposite the deltoid tuberosity

Action: Flexion and adduction of the humerus

Innervation: Musculocutaneous nerve (C5–C7)

Palpation: Palpate medial to short head of biceps brachii when humerus is flexed against resistance.

The **trigger point** for this muscle is near the coracoid attachment, and its **referred pain pattern** is down the triceps and dorsal forearm into the hand. **Synergists** for **adduction** are the **anterior deltoid, pectoralis major**, and the **latissimus dorsi**; for flexion, the **clavicular division of the pectoralis major, anterior deltoid**, and the **biceps brachii. Antagonists** to adduction are the **supraspinatus, infraspinatus**, and the **middle and posterior deltoids**; to flexion, the **posterior deltoid, teres major, latissimus dorsi, triceps brachii (long head)**, and the **pectoralis major (sternal head)**.

Biceps Brachii
(bi•seps)(bray•kee•eye)

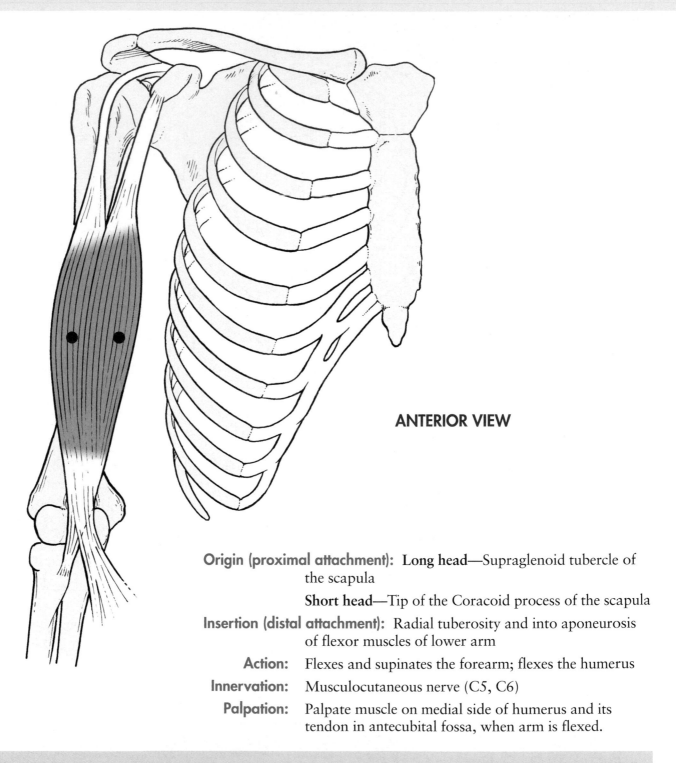

ANTERIOR VIEW

Origin (proximal attachment): **Long head**—Supraglenoid tubercle of the scapula

Short head—Tip of the Coracoid process of the scapula

Insertion (distal attachment): Radial tuberosity and into aponeurosis of flexor muscles of lower arm

Action: Flexes and supinates the forearm; flexes the humerus

Innervation: Musculocutaneous nerve (C5, C6)

Palpation: Palpate muscle on medial side of humerus and its tendon in antecubital fossa, when arm is flexed.

The number of heads may be as high as five, with the additional heads arising from various points along the length of the humerus. The **trigger points** are found in the belly of each of the two heads of the muscle. Its **referred pain pattern** is found in the front of the shoulder and into the antecubital space (the "crease" of the elbow). **Synergists** are the **brachialis, brachioradialis,** and the **supinator; antagonists** are the **triceps brachii** and **anconeus.**

Brachialis
(bray•kee•al•is)

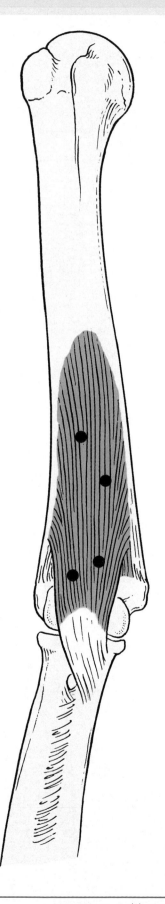

ANTERIOR VIEW

Origin (proximal attachment): Distal half of the anterior surface of the humerus

Insertion (distal attachment): Coronoid process and tuberosity of ulna

Action: Flexes elbow

Innervation: Musculocutaneous and radial nerves (C5, C6)

Palpation: Palpate muscle medial to biceps brachii on lower anterior humerus and its tendon medial to biceps tendon in the antecubital fossa during active flexion of elbow.

The inferior attachment may be divided into two or more parts, inserting on the bones and some muscles of the lower arm. The **trigger points** are found in the belly of the muscle. Its **referred pain pattern** is primarily in the lower arm to the thumb. This muscle is a strong elbow flexor. **Synergists** are the **biceps brachii** and **brachioradialis**; antagonists are the **triceps brachii** and **anconeus**.

Brachioradialis
(bray•kee•oh•ray•de•al•is)

LATERAL VIEW

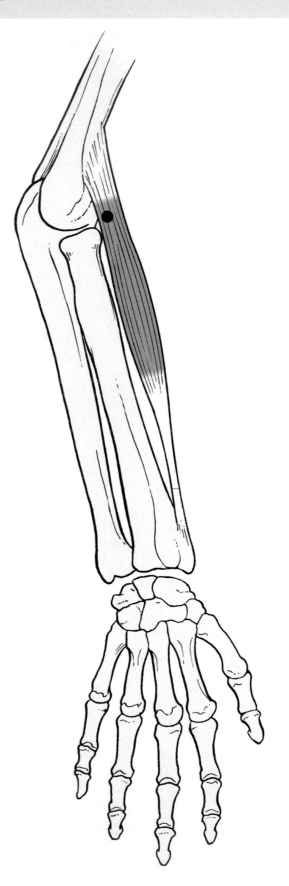

Origin (proximal attachment): Proximal two-thirds of the lateral supracondylar ridge of humerus

Insertion (distal attachment): Lateral side of the base of the styloid process of radius

Action: Flexes the elbow; assists in pronation and supination of the forearm to the midposition

Innervation: Radial nerve (C5, C6)

Palpation: Palpate on upper forearm during resisted elbow flexion.

The **brachioradialis** is used to stabilize the elbow during rapid flexion and extension while in a midposition, such as in hammering. The **trigger point** is in the belly of the muscle. The **referred pain pattern** is from the wrist and base of the thumb in the web space between the thumb and index finger to the lateral epicondyle of the humerus. **Synergists** are the **brachialis** and **biceps brachii**; antagonists are the **triceps brachii** and **anconeus**.

Triceps Brachii
(try•seps)(bray•kee•eye)

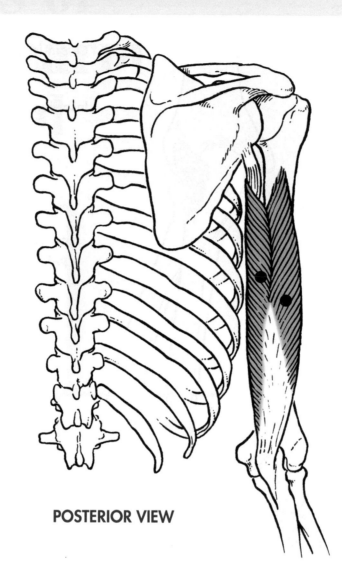

POSTERIOR VIEW

Origin (proximal attachment): **Long head**—infraglenoid tubercle of the scapula

Medial head—distal two-thirds of the medial and posterior surfaces of the humerus

Lateral head—upper half of the posterior surface of the humerus

Insertion (distal attachment): Posterior surface of the olecranon process of the ulna

Action: Extends the forearm and the tendon of the long head helps stabilize the shoulder joint and extends the humerus.

Innervation: Radial nerve (C7, C8)

Palpation: Palpate on dorsal upper arm during active elbow extension.

The triceps brachii is the only muscle on the posterior upper arm. Its **trigger points** are in the belly of each head of the muscle. Its **referred pain pattern** is the entire length of the posterior surface of the arm. It synergist is the **anconeus; antagonists** are the **biceps brachii** and **brachialis.**

Pronator Teres
(pro•**nay**•ter) (**ter**•eez)

ANTERIOR VIEW

Origin (proximal attachment): **Humeral head**—just above the medial epicondyle of the humerus

Ulnar head—medial side of the coronoid process of the ulna

Insertion (distal attachment): Middle of lateral surface of radius

Action: Pronates the forearm and assists in flexing the elbow joint

Innervation: Median nerve (C6, C7)

Palpation: Palpate on medial side of forearm just medial to insertion of biceps brachii during resisted pronation.

This muscle forms the medial border of the antecubital fossa. Its **trigger point** is the belly of the muscle near the elbow attachment. Its **referred pain pattern** is the radial side of the forearm into the wrist and thumb. Its **synergist** is the **pronator quadratus**; its **antagonist** is the **supinator**.

Supinator
(su•pih•nay•ter)

Origin (proximal attachment): Lateral epicondyle of humerus, annular and radial collateral ligaments, and supinator crest of ulna

Insertion (distal attachment): Lateral surface of the upper one third of the body of the radius

Action: Supinates the forearm

Innervation: Radial nerve (C6)

Palpation: Palpate distal to lateral epicondyle of humerus on posterior lateral side of forearm during resisted supination of forearm. Deep muscle; difficult to palpate.

The **supinator** is a large muscle that wraps around the bones of the forearm. This muscle is covered by the more superficial muscles. Its **trigger point** is near the radius in the antecubital space. Its **referred pain pattern** is from the lateral epicondyle to the dorsal web of the thumb. It mimics "**tennis elbow.**" The radial nerve passes between the superficial and deep layers of this muscle. Its **synergists** are the **biceps brachii** and the accessory **supinator** when present; **antagonists** are the **pronator quadratus** and **pronator teres.** An **accessory supinator** may arise from the coronoid process of the ulna just lateral to the ulnar tuberosity, and insert onto the radial tuberosity, posterior to the insertion of the biceps tendon.

Flexor Carpi Radialis
(flex•er) (kar•pee)(ray•de•al•is)

ANTERIOR VIEW

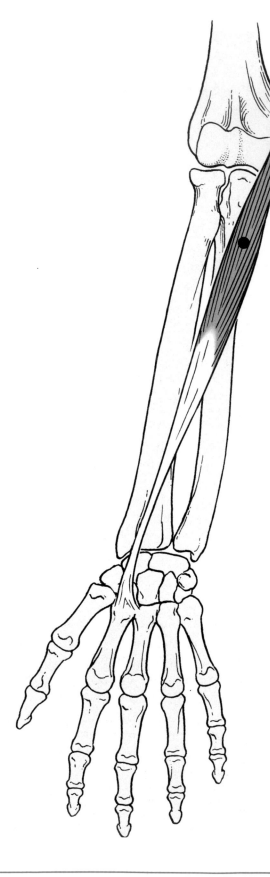

Origin (proximal attachment): Medial epicondyle of the humerus

Insertion (distal attachment): Base of second and third metacarpal bones

Action: Flexes wrist and abducts hand

Innervation: Median nerve (C6, C7)

Palpation: Palpate tendon on anterior surface of wrist in line with second metacarpal.

The **flexor carpi radialis** runs diagonally across the forearm. Its fleshy belly is replaced midway by a flat tendon that becomes cord-like at the wrist. The tendon acts as a guide to the position of the radial artery in order to take a pulse. "Golfer's elbow" is a painful condition that follows repetitive use of the superficial muscles on the anterior forearm, straining the **common flexor tendon**. The **flexor carpi radialis** is the most commonly affected muscle. Another name for this syndrome is **medial epicondylitis**, reflecting the site of inflammation. The **trigger point** is in the belly of the muscle. **Synergists** for flexion are the **flexor carpi ulnaris** and **palmaris longus**; **antagonists** are the **extensor carpi radialis longus** and **brevis**, and **extensor ulnaris**.

Palmaris Longus
(pal•mar•is) (lon•gus)

ANTERIOR VIEW

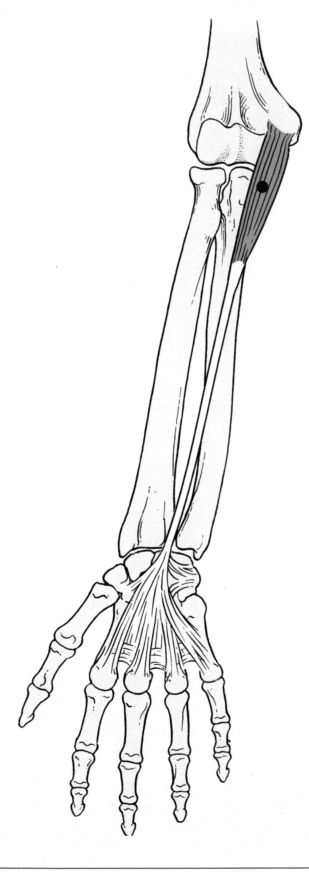

Origin (proximal attachment): Medial epicondyle of the humerus through the common flexor tendon

Insertion (distal attachment): Front of flexor retinaculum and apex of the palmar aponeurosis

Action: Tenses the palmar fascia and flexes the wrist

Innervation: Median nerve (C6–C8)

Palpation: Palpate in midline of anterior surface of wrist when wrist is flexed against resistance and thumb is abducted.

The tendon of the **palmaris longus** is above the antebrachial fascia of the wrist and can be seen if one cups the hand and flexes the wrist. This muscle is absent in about one fourth of the population. Its **trigger point** is in the belly of the muscle. The **referred pain pattern** is into the wrist and fingers. **Synergists** are the **flexor carpi radialis, flexor carpi ulnaris,** and **flexor digitorum superficialis**; its antagonists are the **extensor carpi radialis longus, extensor carpi radialis brevis, extensor carpi ulnaris,** and **extensor digitorum.**

Flexor Carpi Ulnaris
(flex•er) (kar•pee) (ul•nar•is)

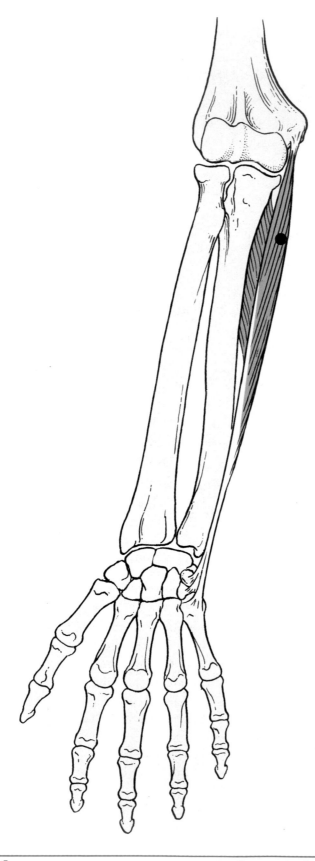

Origin (proximal attachment): **Humeral head—**common tendon from the medial epicondyle of humerus

Ulnar head—olecranon process and proximal two thirds of the posterior border of the ulna

Insertion (distal attachment): Pisiform bone, hook of the hamate, and the base of the fifth metacarpal bone

Action: Flexes and adducts the wrist

Innervation: Ulnar nerve (C7, C8)

Palpation: Palpate tendon on anterior surface of wrist proximal to pisiform bone during active wrist flexion.

The ulnar nerve passes through this muscle. Its **trigger point** is in its belly and its **referred pain pattern** is in the wrist and hand. **Synergists** for flexion are the **flexor carpi ulnaris** and **palmaris longus;** antagonists are the **extensor carpi radialis longus** and **brevis** and the **extensor ulnaris.**

Flexor Digitorum Superficialis
(flex•er) (dij•ih•tor•um)(su•per•fish•ee•al•lis)

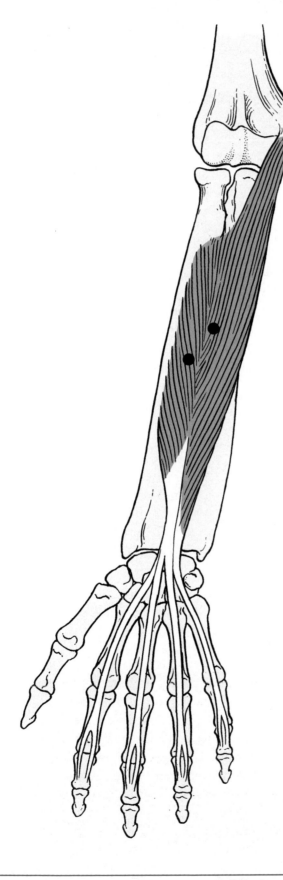

ANTERIOR VIEW

Origin (proximal attachment): **Humeral head**— medial epicondyle of the humerus through the common tendon and the medial margin of the coronoid process of the ulna

Radial head—anterior surface of the shaft of the radius

Insertion (distal attachment): Four tendons divide into two slips each; the slips insert into the sides of the middle phalanges of the four fingers

Action: Flexes the wrist and the middle phalanges of fingers two through five

Innervation: Median nerve (C7–T1)

Palpation: Palpate on anterior surface of wrist between tendons of flexor carpi ulnaris and palmaris longus during active flexion.

The median nerve and ulnar artery pass beneath the origin of this muscle. At the wrist, the tendons of fingers three and four are superficial to the tendons of fingers two and five. Each of the tendons divide at the proximal phalanx to allow the tendon of the flexor digitorum profundus to pass through. The **trigger points** are in the belly of each head of the muscle. Its **synergist** is the **flexor digitorum profundus**; its **antagonist** is the **extensor digitorum**.

Flexor Digitorum Profundus
(flex•er) (dij•ih•tor•um) (pro•fun•dus)

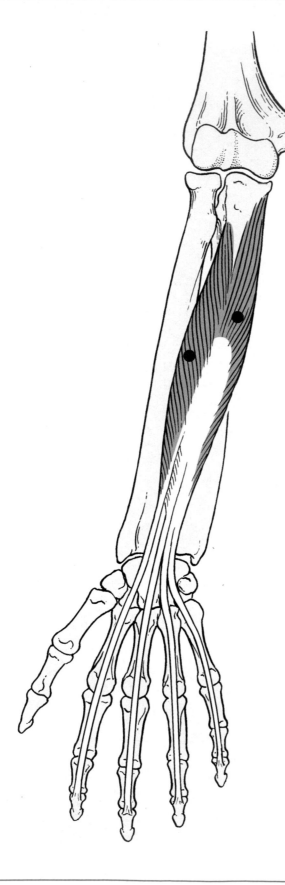

ANTERIOR VIEW

Origin (proximal attachment): Medial and anterior surfaces of the proximal three fourths of the ulna and the interosseous membrane

Insertion (distal attachment): By four tendons into the anterior surface of the distal phalanges of digits two through five

Action: Flexes the distal interphalangeal joints of digits two through five and assists in the adduction of the index, ring, and little fingers and in flexion at the wrist

Innervation: Ulnar nerve and interosseous branch of the median nerve (C7–T1)

Palpation: Palpate tendon on anterior surface of middle phalanges of fingers during active flexion.

An additional radial origin may be found along the entire border of the **flexor pollicis longus.** The tendons of this muscle pass through the tendons of the **flexor digitorum superficialis.** The **trigger points** are in the belly of the muscle. Its **referred pain pattern** is the wrist into the fingers. Its **synergist** is the **flexor digitorum superficialis;** its **antagonist** is the **extensor digitorum.**

Flexor Pollicis Longus
(flex•er) (pohl•ih•sis) (lon•gus)

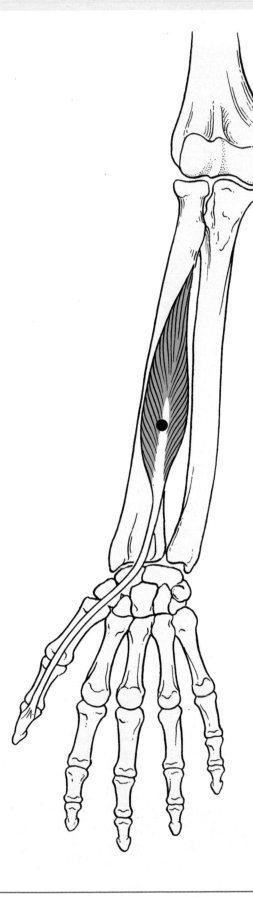

ANTERIOR VIEW

Origin (proximal attachment): Middle of anterior shaft of the radius and interosseous membrane

Insertion (distal attachment): Palmar surface of the base of the distal phalanx of the thumb

Action: Flexes the thumb; assists in abduction of wrist

Innervation: Anterior interosseous branch of the median nerve (C8, T1)

Palpation: Palpate tendon on anterior surface of proximal phalanx of thumb during active flexion.

Additional superior attachments may be from the medial epicondyle and either the medial or lateral side of the coronoid process of the ulna. The **trigger point** is in the belly of the muscle. Its **referred pain pattern** is in the wrist and thumb. **Synergists** are the **flexor pollicis brevis** and the **adductor pollicis;** antagonists are the **extensor pollicis longus, extensor pollicis brevis,** and **abductor pollicis longus.**

Pronator Quadratus
(pro•**nay**•ter) (kwa•**drah**•tus)

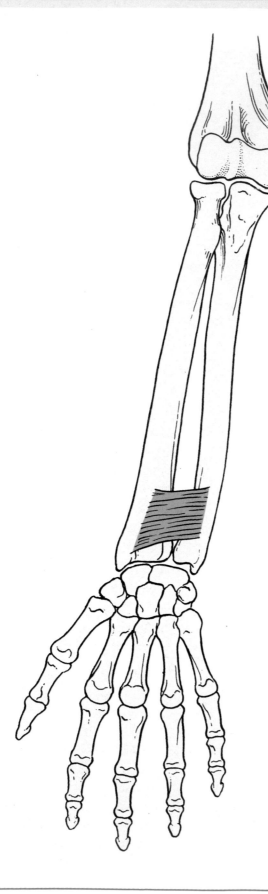

Origin (proximal attachment): Medial side of the anterior surface of the distal one-fourth of the ulna

Insertion (distal attachment): Lateral side of the distal one-fourth of the radius

Action: Pronation of the forearm

Innervation: Anterior interosseous branch of the median nerve (C8, T1)

Palpation: Palpate on lateral anterior forearm on either side of radial pulse.

The **pronator quadratus** is the deepest muscle of the distal forearm. It is the only muscle that arises solely on the ulna and inserts solely on the radius. It is the primary pronator of the forearm. Its **synergist** is the **pronator teres**; its **antagonist** is the **supinator**.

Extensor Carpi Radialis Longus
(ex•**ten**•ser) (**kar**•pee)(**ray**•de•al•is)(**lon**•gus)

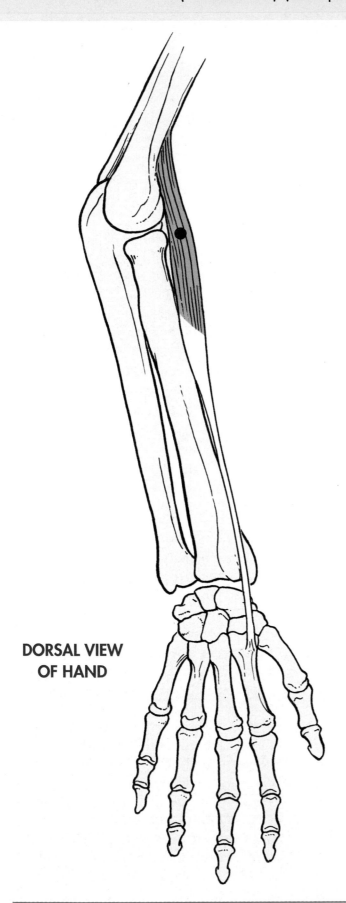

DORSAL VIEW OF HAND

Origin (proximal attachment): Lower third of lateral supracondylar ridge of humerus

Insertion (distal attachment): Dorsal surface of the base of the second metacarpal bone

Action: Extends the wrist and abducts the hand

Innervation: Radial nerve (C6, C7)

Palpation: Palpate tendon on dorsal surface of wrist at base of second metacarpal bone.

It and the **extensor carpi radialis brevis** may be fused with a common belly with two or three insertions to the metacarpal bones. The **trigger point** is in its belly. Its **referred pain pattern** is to the lateral epicondyle and back of hand. This muscle parallels the **brachioradialis** on the lateral forearm. **Synergists** for extension are the **extensor carpi radialis brevis** and **extensor carpi ulnaris**; antagonists are the **flexor carpi radialis, flexor carpi ulnaris**, and **palmaris longus.**

Extensor Carpi Radialis Brevis
(ex•ten•ser) (kar•pee) (ray•de•a•lis) (brev•is)

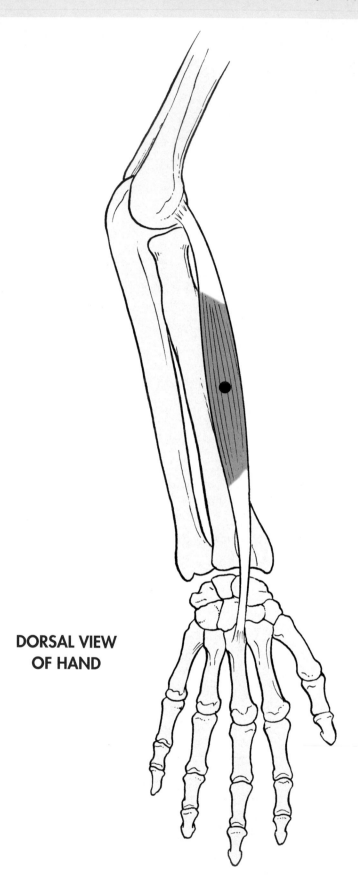

**DORSAL VIEW
OF HAND**

Origin (proximal attachment): Lateral epicondyle of the humerus and the radial collateral ligament

Insertion (distal attachment): Dorsal surface of the base of the third metacarpal bone

Action: Extends the wrist and assists in abduction of the hand

Innervation: Radial nerve (C6, C7)

Palpation: Palpate tendon on dorsal surface of wrist at base of third metacarpal medial to tendon for extensor carpi radialis longus.

The **trigger point** is in the belly of the muscle. Its **referred pain pattern** is from the lateral epicondyle of the humerus down the posterior portion of the fore-arm to the hand and the middle finger. Occasionally, a **cystic swelling** occurs on its tendon, causing a grape-sized bump to appear on the wrist or back of the hand. Sometimes erroneously called a ganglion, it enlarges during flexion. The distal attachment of this muscle's tendon into the base of the third metacarpal bone is a common site. **Synergists** for extension are the **extensor carpi radialis longus** and **extensor carpi ulnaris**; **antagonists** are the **flexor carpi radialis, flexor carpi ulnaris**, and **palmaris longus**.

Extensor Digitorum
(ex•**ten**•ser) (dij•ih•**tor**•um)

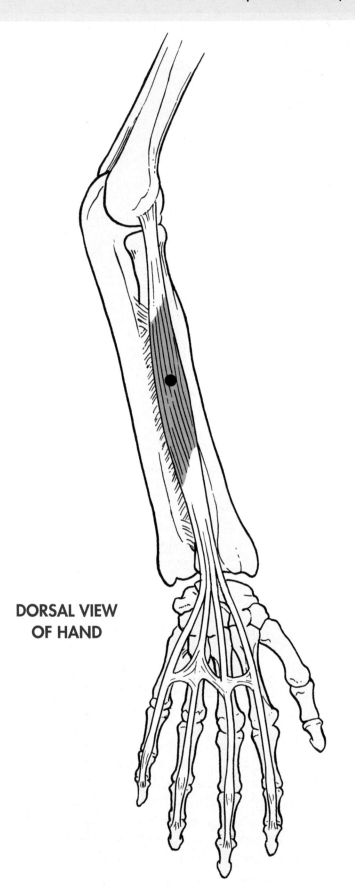

**DORSAL VIEW
OF HAND**

Origin (proximal attachment): Common extensor tendon from the lateral epicondyle of humerus

Insertion (distal attachment): By four tendons to the lateral and dorsal surfaces of all the phalanges of digits two through five

Action: Extends the fingers and the wrist

Innervation: Deep branch of radial nerve (C6–C8)

Palpation: Palpate muscle on middle of dorsal forearm during forced finger and wrist extension; palpate its tendons on dorsal surface of hand.

This muscle divides into four prominent tendons on the dorsum of the hand. Sudden tension on a long extensor tendon may tear its attachment to a finger at the distal phalangeal joint. This occurs when the finger is forced into extreme flexion. This is called "**mallet finger**" or baseball finger. This muscle tends to hyperextend the metacarpophalangeal joint. Inflammation of this common tendon of origin is a primary cause of **tennis elbow**. The **trigger point** is in the belly of the muscle. **Synergists** are the **extensor indicis, extensor digiti minimi,** and the **lumbricales; antagonists** are the **flexor digitorum superficialis,** and the **flexor digitorum profundus.**

Extensor Carpi Ulnaris
(ex•ten•ser) (kar•pee) (ul•nar•is)

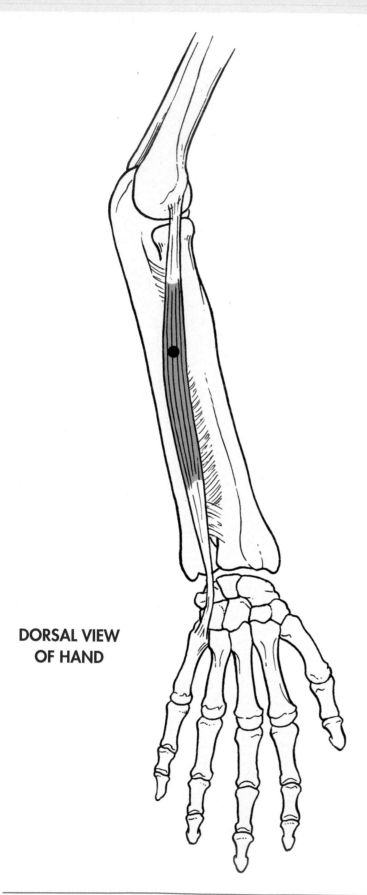

DORSAL VIEW OF HAND

Origin (proximal attachment): Common tendon from the lateral epicondyle of humerus and posterior border of ulna

Insertion (distal attachment): Dorsal surface of base of fifth metacarpal bone

Action: Extends the wrist and assists in adduction of hand

Innervation: Interosseous branch of the radial nerve (C6–C8)

Palpation: Palpate along posterior border of ulna; palpate its tendon on dorsal surface of wrist on ulnar side of carpal bones;

The tendon of the **extensor carpi ulnaris** runs through a groove between the head of the ulna and the styloid process of the ulna. Its **trigger point** is in its belly. Its **referred pain pattern** is from the lateral epicondyle down the posterior surface of the forearm to the little finger. **Synergists** for extension are the **extensor carpi radialis longus** and the **extensor carpi radialis brevis**; **antagonists** are the **flexor carpi radialis, flexor carpi ulnaris,** and **palmaris longus.**

Abductor Pollicis Longus

(ab•**duck**•ter) (**pohl**•ih•sis) (**lon**•gus)

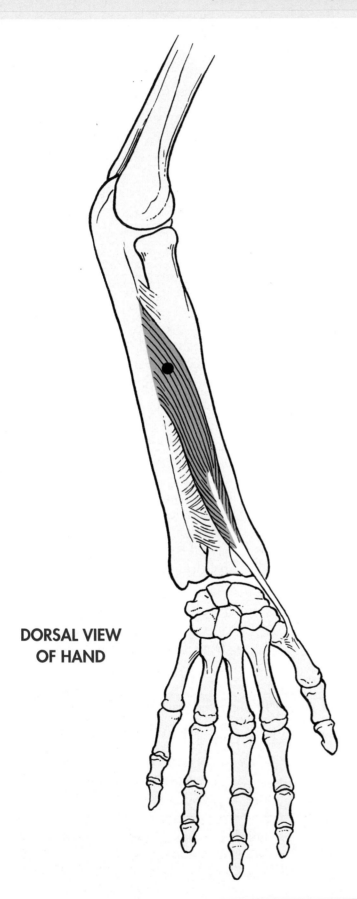

**DORSAL VIEW
OF HAND**

Origin (proximal attachment): Posterior surface of the body of the ulna distal to the origin of the supinator, the interosseous membrane, and the middle one third of the body of the radius

Insertion (distal attachment): Dorsal surface of the base of the first metacarpal bone

Action: Abducts and extends the first digit and abducts the wrist

Innervation: Posterior interosseous branch of the radial nerve (C6–C8)

Palpation: Palpate on radial side of wrist at base of first metacarpal during active abduction of thumb.

The **trigger point** is in the belly of the muscle. Its **referred pain pattern** is from the lateral epicondyle to the web of the thumb. **Synergists** for extension are the **extensor pollicis brevis** and the **abductor pollicis brevis; antagonists** are the **extensor pollicis longus, extensor pollicis brevis,** and **adductor pollicis.**

Extensor Pollicis Longus
(ex•**ten**•ser) (**pohl**•ih•sis) (**lon**•gus)

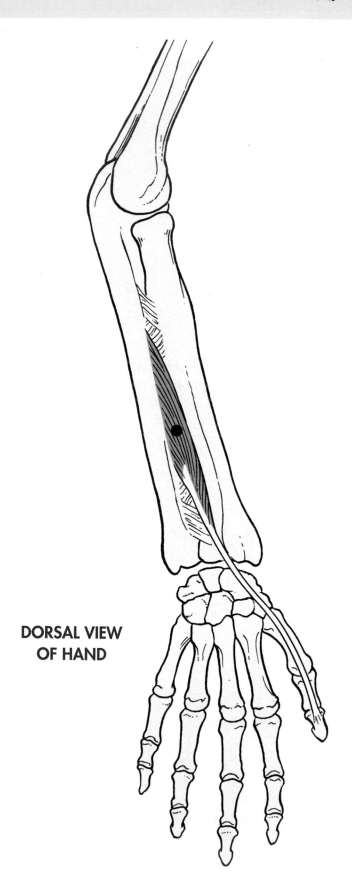

**DORSAL VIEW
OF HAND**

Origin (proximal attachment): Interosseous membrane and middle one third of the posterior surface of the ulna

Insertion (distal attachment): Dorsal surface of the base of the distal phalanx of the first digit

Action: Extends the interphalangeal joint and assists in extension of the metacarpophalangeal joints in the thumb; it also assists in abduction and extension of the wrist and the lateral rotation of the thumb

Innervation: Posterior interosseous branch of the radial nerve (C6–C8)

Palpation: Palpate tendon on ulnar side of dorsal proximal phalanx of thumb during active extension.

This muscle forms the lateral border of the "**anatomical snuffbox**" described with the **extensor pollicis brevis**. Its **trigger point** is in the belly of the muscle. Its **referred pain pattern** is from the lateral epicondyle to the thumb. Its **synergist** is the **extensor pollicis brevis**; its **antagonist** is the **flexor pollicis longus**.

Extensor Pollicis Brevis
(ex•ten•ser) (pohl•ih•sis) (brev•is)

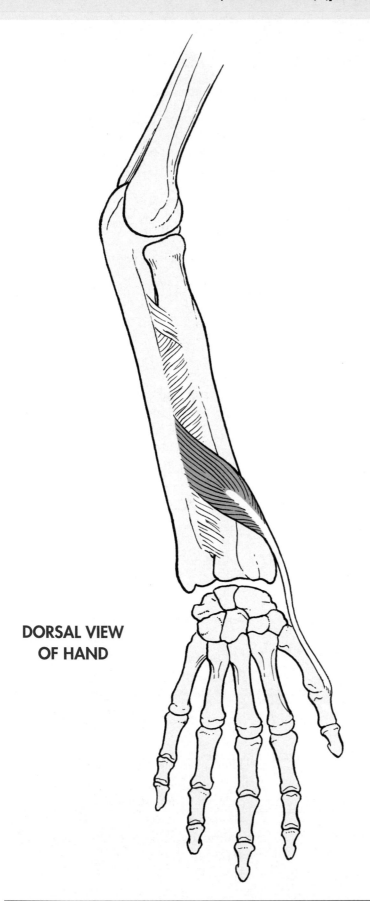

DORSAL VIEW OF HAND

Origin (proximal attachment): Distal one-third of dorsal surface of radius and adjacent part of interosseus membrane

Insertion (distal attachment): Base of proximal phalanx of the thumb

Action: Extends thumb and abducts hand

Innervation: Radial nerve (C6, C7)

Palpation: Palpate tendon during active abduction and extension of thumb.

Synergists are the **abductor pollicis longus** and **extensor pollicis longus**; **antagonists** are the **adductor pollicis** and **flexor pollicis longus**. The **scaphoid bone** forms the bottom of the "anatomical snuffbox" bordered medially by the **extensor pollicis brevis** and laterally by the **extensor pollicis longus**. Pain in response to palpation in this region may indicate a possible fracture of the scaphoid bone.

Extensor Indicis
(ex•**ten**•ser) (**in**•dih•sis)

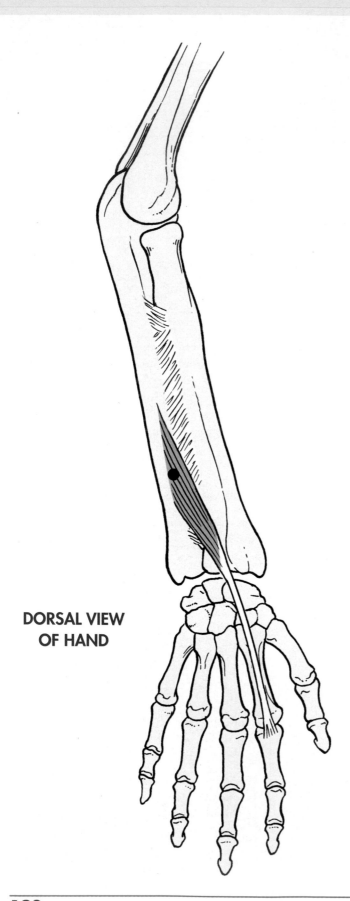

**DORSAL VIEW
OF HAND**

Origin (proximal attachment):
Interosseous membrane
and posterior surface of
the ulna

Insertion (distal attachment): Into
extensor expansion
on dorsal surface of
proximal phalanx of
the index finger

Action: Extends the index finger
at metacarpalphalanx
joint

Innervation: Posterior interosseous
branch of the radial
nerve (C6–C8)

Palpation: Palpate tendon at base of
index finger during
resisted finger extension.

This is a tiny muscle. The **trigger
point** is in the belly of the muscle. Its
synergist is the **extensor digitorum**;
antagonists are the **flexor digitorum
superficialis** and **flexor digitorum
profundus.**

Extensor Digiti Minimi
(ex•**ten**•ser)(**dij**•ih•tye)(**min**•ih•mih)

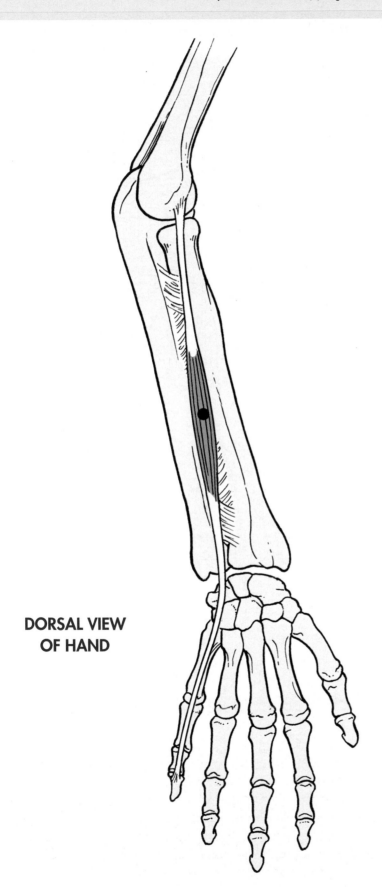

**DORSAL VIEW
OF HAND**

Origin (proximal attachment): Common tendon from the lateral epicondyle of the humerus

Insertion (distal attachment): Dorsal surface of base of the distal phalanx of the fifth digit

Action: Extends the fifth digit

Innervation: Radial nerve (C6–C8)

Palpation: Palpate tendon on ulnar side of tendon of extensor digitorum during resisted extension

The **trigger point** is in the belly of the muscle. Its **synergist** is the **extensor digitorum; antagonists** are the **flexor digitorum superficialis, flexor digitorum profundus,** and **flexor digiti minimi.**

Abductor Pollicis Brevis

(ab•**duck**•ter)(**pohl**•ih•sis) (**brev**•is)

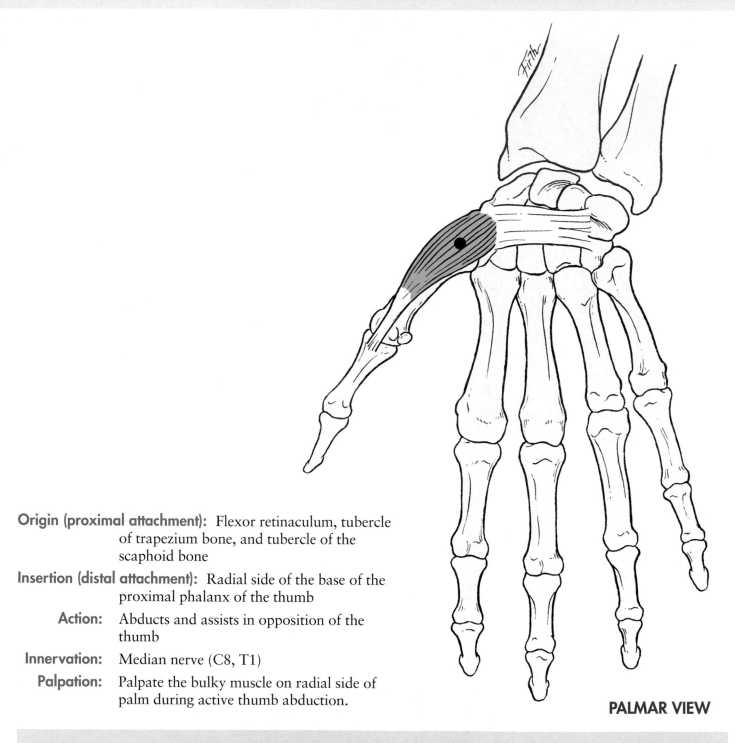

PALMAR VIEW

Origin (proximal attachment): Flexor retinaculum, tubercle of trapezium bone, and tubercle of the scaphoid bone

Insertion (distal attachment): Radial side of the base of the proximal phalanx of the thumb

Action: Abducts and assists in opposition of the thumb

Innervation: Median nerve (C8, T1)

Palpation: Palpate the bulky muscle on radial side of palm during active thumb abduction.

This forms the bulk of the muscle on the radial side of the palm during active thumb abduction. Its **trigger point** is in the belly of the muscle. Its **referred pain pattern** is the wrist into the thumb. Any lesion that reduces the size of the **carpal tunnel (flexor retinaculum)** may cause compression of the **median nerve**. This causes weakness in the **abductor pollicis brevis** and **opponens pollicis** muscles, causing difficulty in performing fine movements with the thumb. **Synergists** are the **abductor pollicis longus** and **extensor pollicis brevis**; **antagonists** are the **flexor pollicis longus, adductor pollicis,** and **opponens pollicis**.

Flexor Pollicis Brevis
(flex•er)(pohl•ih•sis) (brev•is)

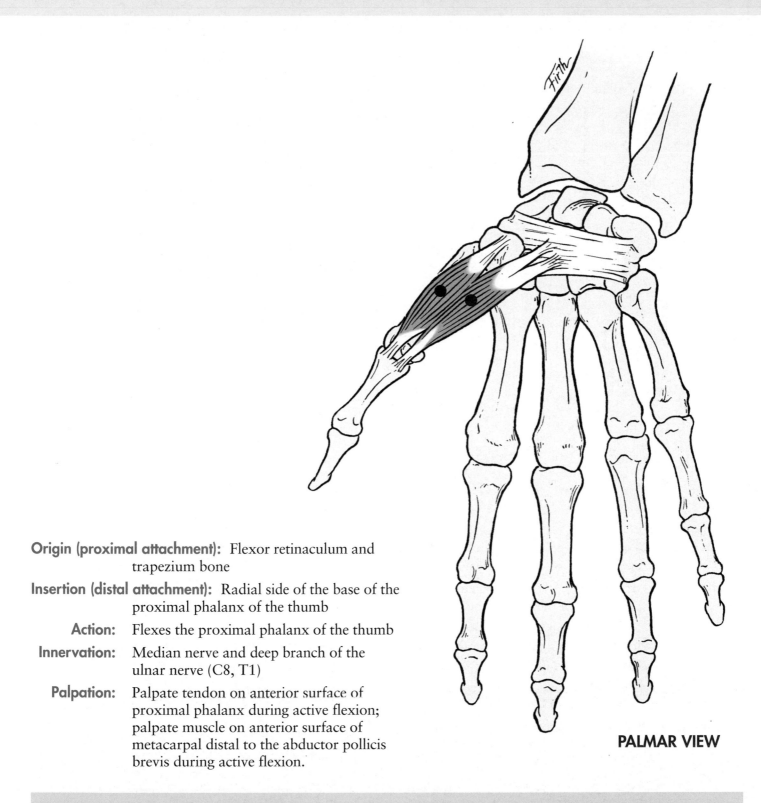

PALMAR VIEW

Origin (proximal attachment): Flexor retinaculum and trapezium bone

Insertion (distal attachment): Radial side of the base of the proximal phalanx of the thumb

Action: Flexes the proximal phalanx of the thumb

Innervation: Median nerve and deep branch of the ulnar nerve (C8, T1)

Palpation: Palpate tendon on anterior surface of proximal phalanx during active flexion; palpate muscle on anterior surface of metacarpal distal to the abductor pollicis brevis during active flexion.

The **trigger points** are in the belly of the muscle. The **referred pain pattern** is from the wrist to the thumb. **Synergists** are the **flexor pollicis longus** and **adductor pollicis**; antagonists are the **extensor pollicis longus, extensor pollicis brevis,** and **abductor pollicis.**

Opponens Pollicis
(op•**po**•nens) (**pohl**•ih•sis)

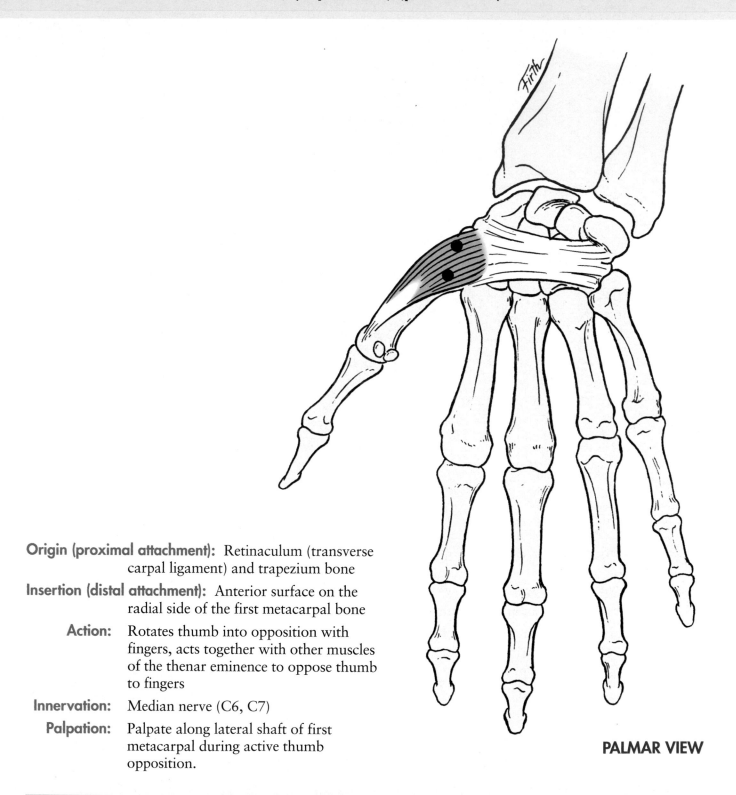

PALMAR VIEW

Origin (proximal attachment): Retinaculum (transverse carpal ligament) and trapezium bone

Insertion (distal attachment): Anterior surface on the radial side of the first metacarpal bone

Action: Rotates thumb into opposition with fingers, acts together with other muscles of the thenar eminence to oppose thumb to fingers

Innervation: Median nerve (C6, C7)

Palpation: Palpate along lateral shaft of first metacarpal during active thumb opposition.

The **trigger points** are in the belly. Its **referred pain pattern** is from the wrist into the thumb. **Synergists** are the **flexors pollicis longus** and **brevis** and the **adductor pollicis;** antagonists are the **extensors pollicis longus** and **brevis**, and the **abductors pollicis longus** and **brevis.**

Adductor Pollicis
(ad•**duck**•ter) (**pohl**•ih•sis)

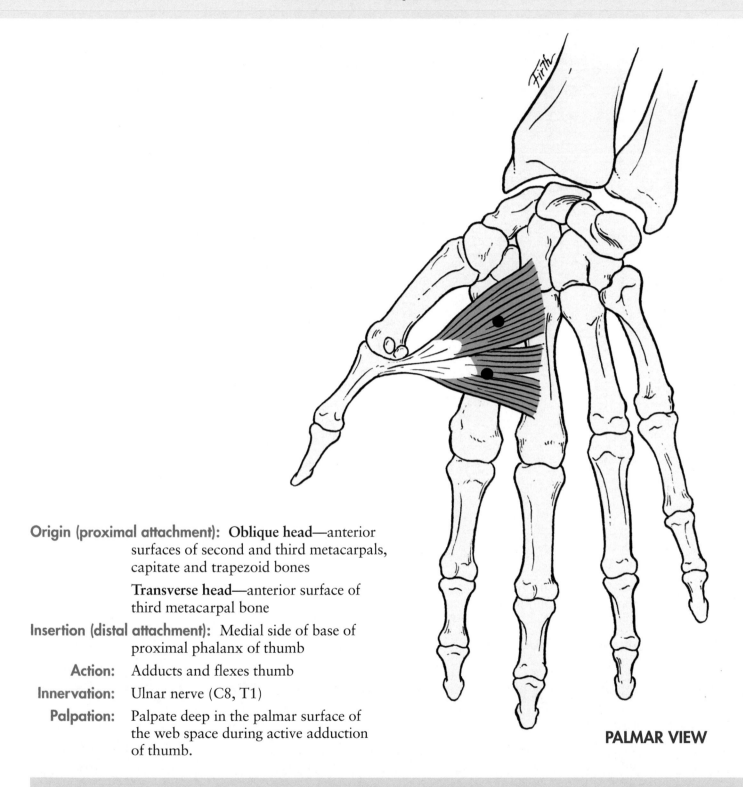

PALMAR VIEW

Origin (proximal attachment): **Oblique head**—anterior surfaces of second and third metacarpals, capitate and trapezoid bones

Transverse head—anterior surface of third metacarpal bone

Insertion (distal attachment): Medial side of base of proximal phalanx of thumb

Action: Adducts and flexes thumb

Innervation: Ulnar nerve (C8, T1)

Palpation: Palpate deep in the palmar surface of the web space during active adduction of thumb.

This fan-shaped muscle forms the bulk of the thumb web space on the anterior surface. Its **trigger points** are in the belly of the muscle. Its **referred pain pattern** is into the thumb. **Synergists** are the **flexor pollicis brevis, flexor pollicis longus,** and **opponens pollicis;** antagonists are the **extensor pollicis brevis, extensor pollicis longus, abductor pollicis longus,** and **abductor pollicis brevis.**

Abductor Digiti Minimi
(ab•duck•ter)(dij•ih•tye)(min•ih•mih)

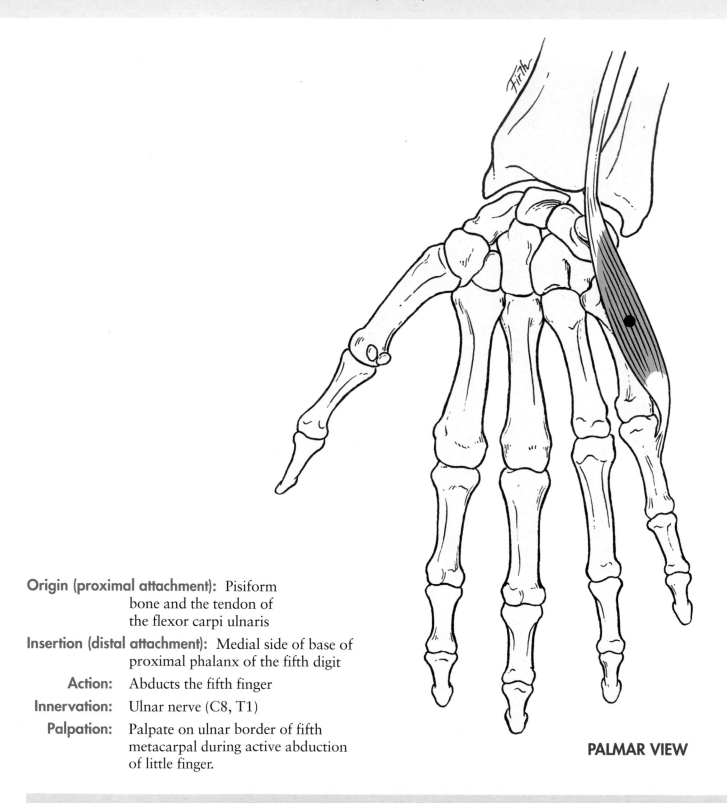

PALMAR VIEW

Origin (proximal attachment): Pisiform bone and the tendon of the flexor carpi ulnaris

Insertion (distal attachment): Medial side of base of proximal phalanx of the fifth digit

Action: Abducts the fifth finger

Innervation: Ulnar nerve (C8, T1)

Palpation: Palpate on ulnar border of fifth metacarpal during active abduction of little finger.

The hypothenar eminence (base of the little finger) is less prominent than the thenar eminence (base of the thumb.) Its **trigger point** is in the belly of the muscle. Its **referred pain pattern** is into the little finger. **Synergists** are the **flexor digiti minimi brevis**, and **opponens digiti minimi**; **antagonists** are the **palmar interossei**.

An Illustrated Atlas of the Skeletal Muscles

Flexor Digiti Minimi Brevis
(**flex**•er)(**dij**•ih•tye)(**min**•ih•mih)(**brev**•is)

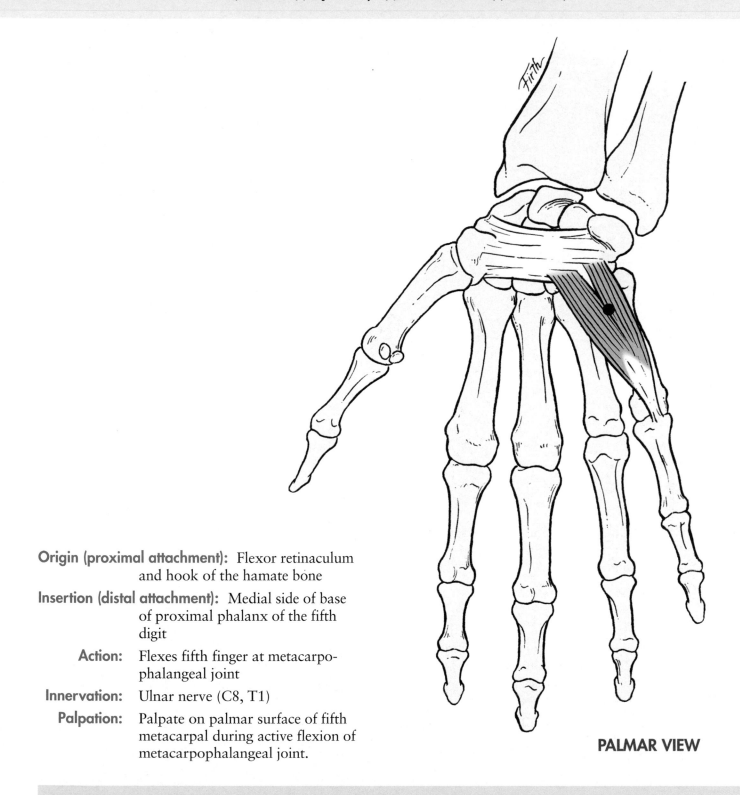

PALMAR VIEW

Origin (proximal attachment): Flexor retinaculum and hook of the hamate bone

Insertion (distal attachment): Medial side of base of proximal phalanx of the fifth digit

Action: Flexes fifth finger at metacarpophalangeal joint

Innervation: Ulnar nerve (C8, T1)

Palpation: Palpate on palmar surface of fifth metacarpal during active flexion of metacarpophalangeal joint.

Either head or the entire muscle may be absent. Its **trigger point** is in its belly and its **referred pain pattern** is into the little finger. **Synergists** are the **flexor digitorum superficialis, flexor digitorum profundus,** and the **opponens digiti minimi;** its **antagonist** is the **extensor digitorum.**

Opponens Digiti Minimi
(op•**po**•nens)(**dij**•ih•tye)(**min**•ih•mih)

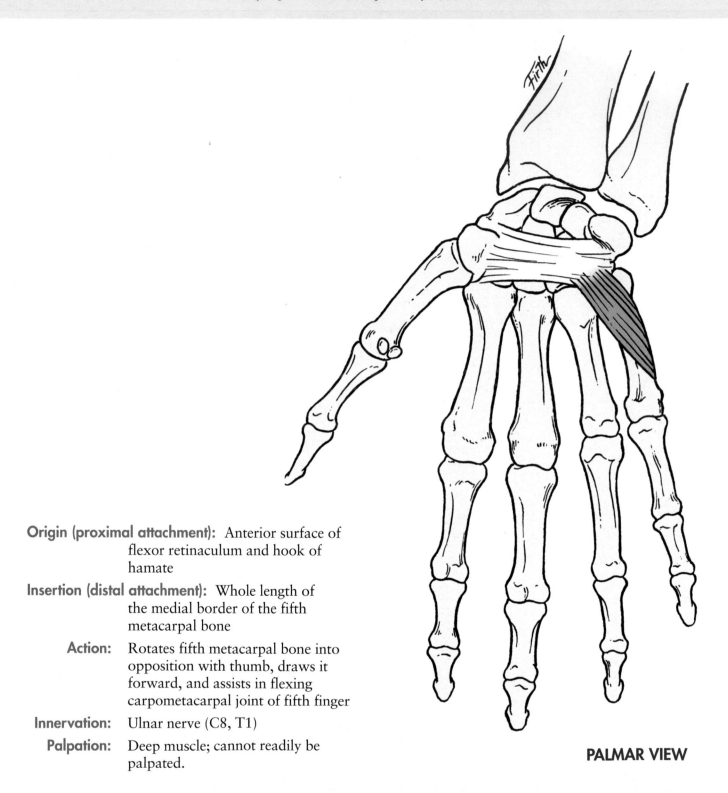

PALMAR VIEW

Origin (proximal attachment): Anterior surface of flexor retinaculum and hook of hamate

Insertion (distal attachment): Whole length of the medial border of the fifth metacarpal bone

Action: Rotates fifth metacarpal bone into opposition with thumb, draws it forward, and assists in flexing carpometacarpal joint of fifth finger

Innervation: Ulnar nerve (C8, T1)

Palpation: Deep muscle; cannot readily be palpated.

This muscle helps to cup the palm of the hand. It lies deep to the **abductor digiti minimi** and **flexor digiti minimi**. **Synergists** are the **flexor digiti minimi brevis** and the **abductor digiti minimi**; **antagonists** are the **dorsal interossei**.

Palmaris Brevis
(pal•mar•is) (brev•is)

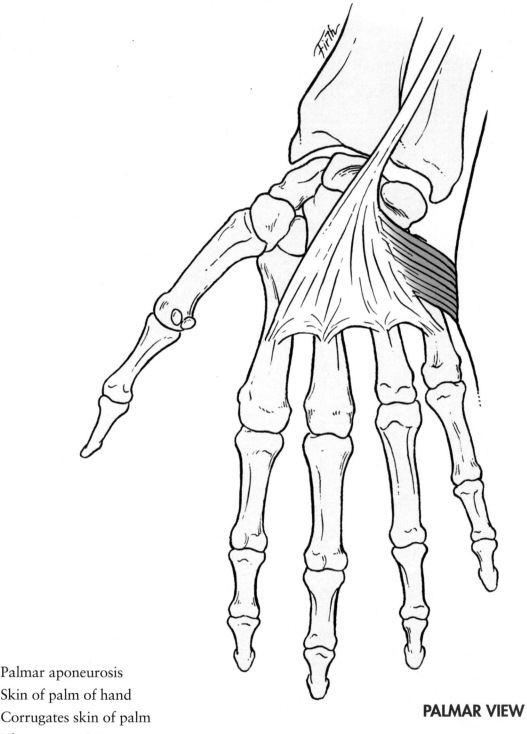

PALMAR VIEW

Origin:	Palmar aponeurosis
Insertion:	Skin of palm of hand
Action:	Corrugates skin of palm
Innervation:	Ulnar nerve (C8)
Palpation:	Cannot readily be palpated.

This is a small muscle lying in the **fascia** of the **hypothenar eminence**. It is often absent.

Lumbricales—Lumbricals
(lum•brih•kay•leez)

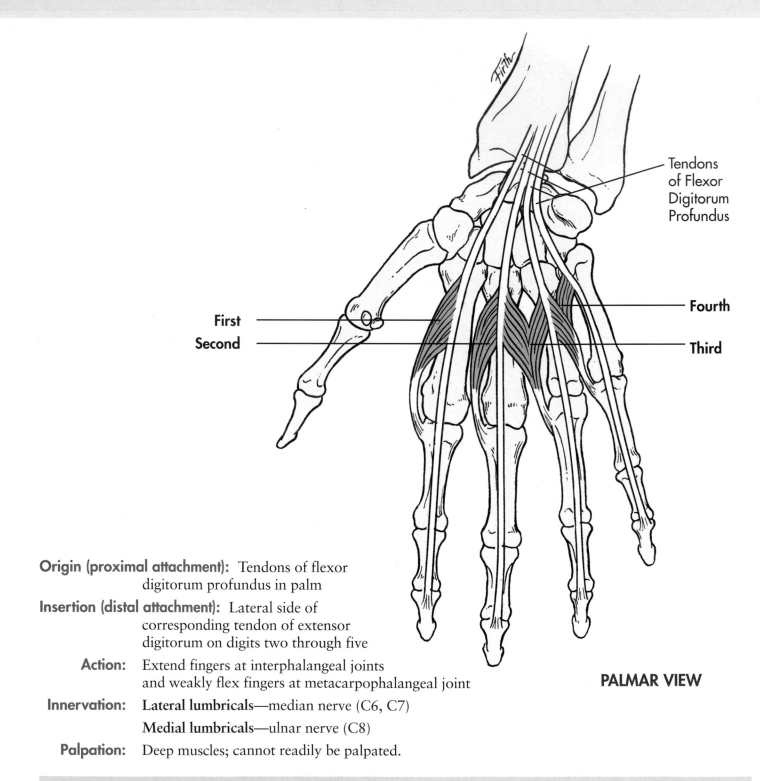

Tendons
of Flexor
Digitorum
Profundus

First

Second

Fourth

Third

PALMAR VIEW

Origin (proximal attachment): Tendons of flexor digitorum profundus in palm

Insertion (distal attachment): Lateral side of corresponding tendon of extensor digitorum on digits two through five

Action: Extend fingers at interphalangeal joints and weakly flex fingers at metacarpophalangeal joint

Innervation: **Lateral lumbricals**—median nerve (C6, C7)

Medial lumbricals—ulnar nerve (C8)

Palpation: Deep muscles; cannot readily be palpated.

These four muscles assist the **extensor digitorum** in extending the fingers. Simultaneous flexion at the metocarpophalangeal joint and extension at the interphalangeal joints as in holding a cup or pencil is characteristic of these muscles. **Synergists** are the **extensor digitorum, extensor indices** and the **extensor digiti minimi**; **antagonists** are the **flexors digitorum superficialis** and **profundus**.

Palmar Interossei
(pal•**mar**)(in•ter•**os**•ee•eye)

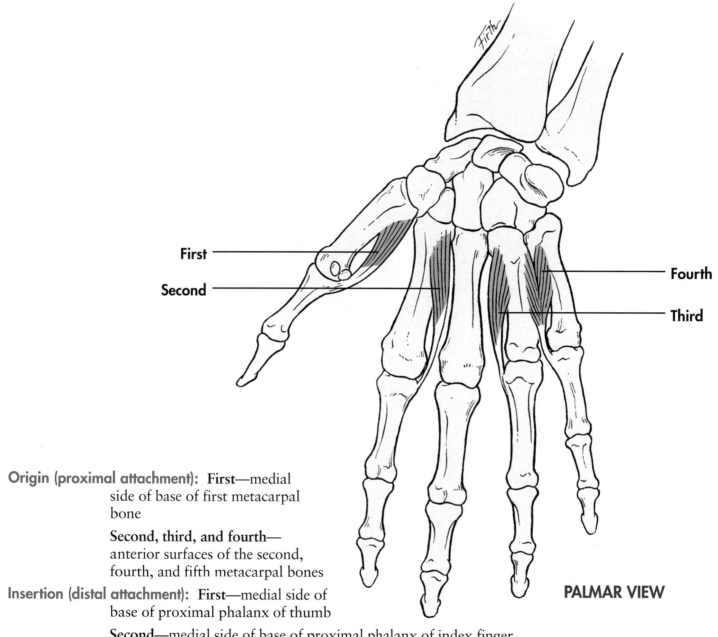

First

Second

Fourth

Third

Fifth

PALMAR VIEW

Origin (proximal attachment): **First**—medial side of base of first metacarpal bone

Second, third, and fourth—anterior surfaces of the second, fourth, and fifth metacarpal bones

Insertion (distal attachment): **First**—medial side of base of proximal phalanx of thumb

Second—medial side of base of proximal phalanx of index finger

Third and fourth—lateral side of proximal phalanges of ring finger and fifth finger

Action: Adducts fingers toward center of third finger at metacarpophalangeal joints and assist in flexion of fingers at the same joints

Innervation: Deep branch of ulnar nerve (C7–T1)

Palpation: Deep muscles; cannot readily be palpated.

Synergists are the **flexors digitorum superficialis** and **profundus** and the **flexor digiti minimi;** antagonists are the **dorsal interossei, abductor digiti minimi,** and the **lumbricales.**

Dorsal Interossei
(dor•sal)(in•ter•oss•ee•eye)

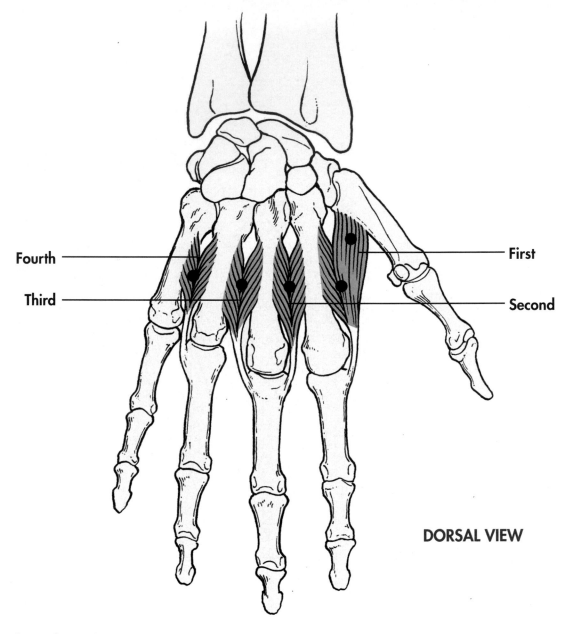

Fourth

Third

First

Second

DORSAL VIEW

Origin (proximal attachment): Adjacent sides of every two metacarpal bones

Insertion (distal attachment): Radial side of the proximal phalanx of the second digit, radial and ulnar side of the third digit, and ulnar side of the fourth digit.

Action: Abducts the index, middle, and ring fingers from the midline of the hand.

Innervation: Deep branch of the ulnar nerve (C8, T1)

Palpation: Palpate along dorsal side of metacarpals during active abduction of fingers.

The **trigger points** are in the belly of the muscles. **Synergists** are the **abductor digiti minimi, extensor digitorum, extensor digiti minimi** and the **lumbricales**; antagonists are the **palmar interossei, flexors digitorum superficialis** and **profundus.**

Muscles of the Hip and Thigh

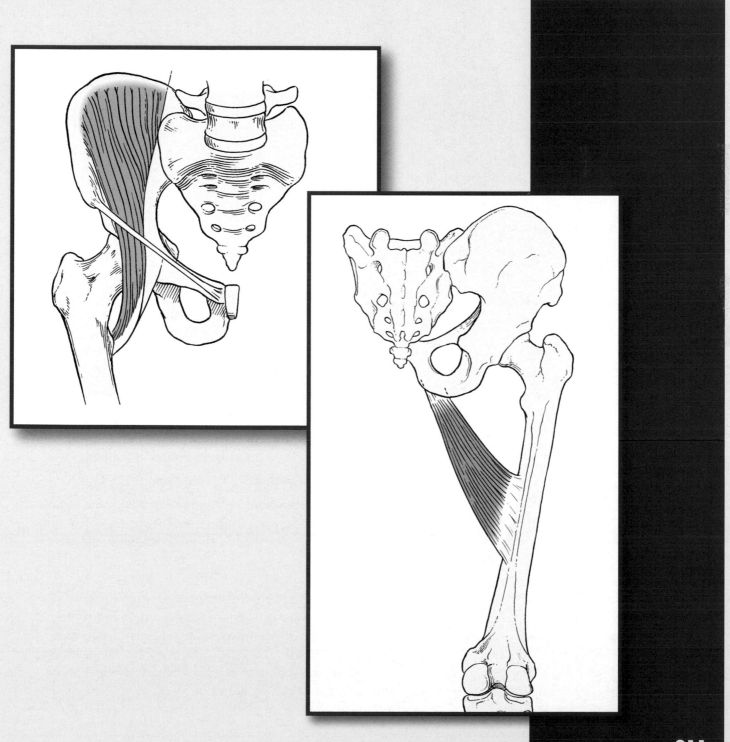

Psoas Major
(so•ahs)

ANTERIOR VIEW

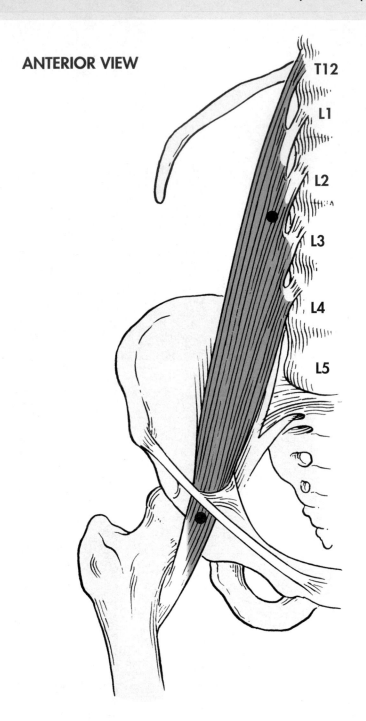

T12
L1
L2
L3
L4
L5

Origin (superior attachment): Transverse processes of all lumbar vertebrae, bodies of last thoracic and all lumbar vertebrae, and intervertebral disk of each lumbar vertebrae

Insertion (inferior attachment): Lesser trochanter of femur

Action: Flexes thigh at the hip joint and flexes vertebral column

Innervation: Ventral rami of L2–L4

Palpation: Palpate deeply between umbilicus and anterior superior iliac spine with thigh flexed at hip joint.

This muscle together with the **iliacus** makes up the **iliopsoas**. Its **trigger points** are near both points of attachment. Its **referred pain pattern** is the entire lumbar area. **Synergists** are the **iliacus**, **adductor group**, and **rectus femoris**; antagonists are the **gluteus maximus** and **hamstring muscles**. In some people, a **psoas minor muscle** may occur medially to the p. major, either unilaterally or bilaterally. It may arise from either the body of T12 or L1 and the disk between them and may insert on the iliac fascia, inguinal ligament, or on the neck or lesser trochanter of the femur. It is **synergistic** with the **psoas major**.

Iliacus
(ill•ee•ak•us)

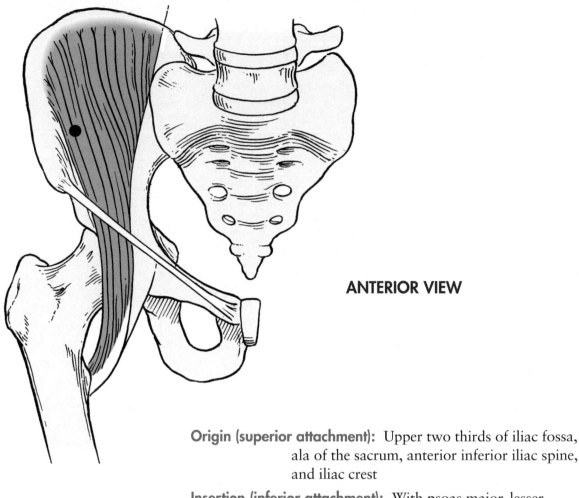

ANTERIOR VIEW

Origin (superior attachment): Upper two thirds of iliac fossa, ala of the sacrum, anterior inferior iliac spine, and iliac crest

Insertion (inferior attachment): With psoas major, lesser trochanter of femur

Action: Flexes thigh at hip joint

Innervation: Muscular branches of femoral nerve (L2–L4)

Palpation: Palpate into the iliac fossa with fingertips while thigh is flexed at hip joint. Most of Iliacus is not palpable.

Contraction of this large fan-shaped muscle brings the swinging leg forward in walking or running. Its **trigger point** is near the inner border of the ilium behind the anterior inferior iliac spine. Its **referred pain pattern** is the entire lumbar area and front of thigh. It can mimic menstrual pain and appendicitis. The combined iliacus/psoas (iliopsoas) muscle and the adductor group of thigh muscles are the muscles usually involved in the injury called a **"pulled groin"** or a **"groin strain,"** since the superior attachments of these muscles are in the inguinal (groin) region at the junction of the abdomen and thigh. Sports involving quick starts such as short-distance racing, basketball, football, and soccer, often produce this tearing within the muscle and at its attachments. **Synergists** are the **psoas major, adductor group,** and **rectus femoris; antagonists** are the **gluteus maximus** and three **hamstring muscles.**

Piriformis
(pih•rih•**for**•mis)

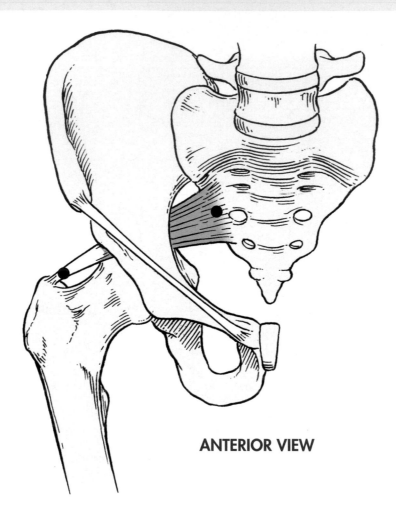

ANTERIOR VIEW

Origin (proximal attachment): Pelvic surface of the sacrum between the first through fourth sacral foramina and sacrotuberous ligament

Insertion (distal attachment): Superior border of the greater trochanter of the femur

Action: Laterally rotates thigh at the hip joint and abducts thigh

Innervation: Anterior rami of S1, S2

Palpation: Palpate half way between posterior superior iliac spine and sacrum during active lateral rotation of hip.

The **trigger points** are near the points of attachment. Its **referred pain pattern** is in the sacroiliac region, the entire buttock and down the posterior thigh. Tension in this muscle may cause entrapment of the sciatic nerve that normally passes under the **piriformis** but which in some individuals may pass through the muscle. **Synergists** are the **superior** and **inferior gemelli, quadratus femoris,** and **internal** and **external obturators; antagonist** is the **gluteus minimus.**

Obturator Externus

(ob•too•**ray**•ter) (ex•**ter**•nus)

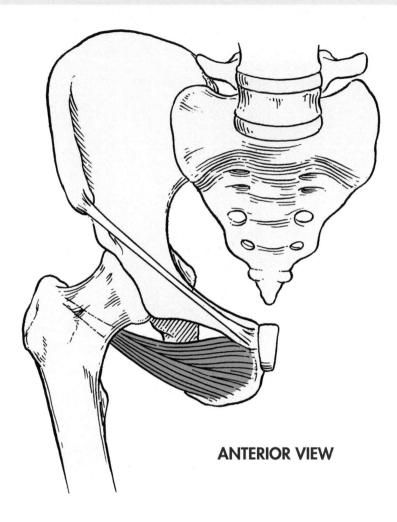

ANTERIOR VIEW

Origin (proximal attachment): Outer surface of superior and inferior rami of pubis and ramus of ischium

Insertion (distal attachment): Trochanteric fossa of femur

Action: Laterally rotates thigh at the hip joint

Innervation: Obturator nerve (L3, L4)

Palpation: Palpate slightly lateral to superior margin of ischial tuberosity during active lateral rotation of thigh at hip joint.

This is a flat triangular muscle deep in the upper medial aspect of the hip under the gluteal muscles. **Synergists** are the **superior** and **inferior gemelli, quadratus femoris, internal obturator,** and **piriformis**; its **antagonist** is the **gluteus minimus.**

Obturator Internus

(ob•too•**ray**•ter) (in•**ter**•nus)

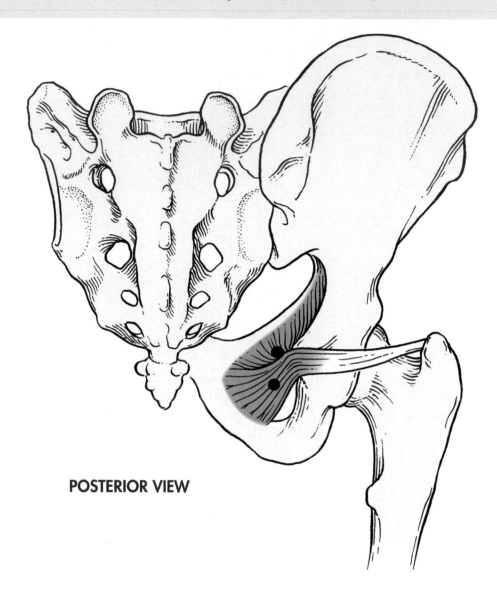

POSTERIOR VIEW

Origin (proximal attachment): Pelvic surface of the obturator membrane and the margins of the obturator foramen; also the internal surface of the pubis and ramus of the ischium

Insertion (distal attachment): Medial surface of the greater trochanter of the femur

Action: Laterally rotates thigh at hip joint

Innervation: L5–S2

Palpation: Deep muscle; cannot be palpated.

This muscle surrounds the **obturator foramen** in the pelvis. It leaves the pelvis by the lesser sciatic notch and turns sharply forward to insert on the greater trochanter. **Trigger points** are in the belly of the muscle. **Synergists** are the **superior** and **inferior gemelli**, **quadratus femoris**, **piriformis**, and **external obturator**; its **antagonist** is the **gluteus minimus**.

Gemellus Superior
(jee•**mel**•lus)

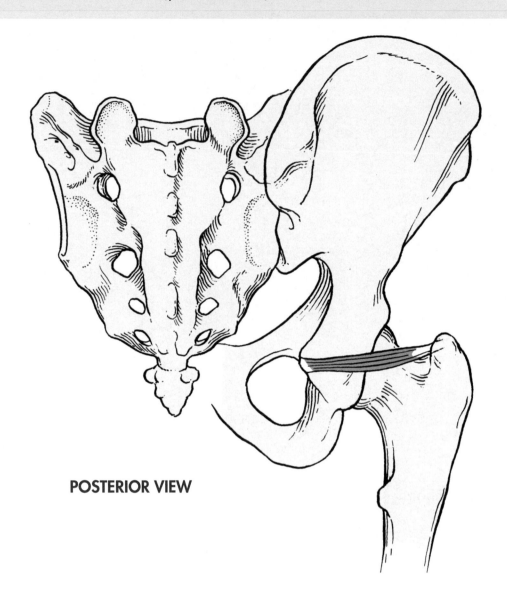

POSTERIOR VIEW

Origin (proximal attachment): Dorsal surface of ischial spine

Insertion (distal attachment): With tendon of obturator internus into the upper border of the greater trochanter

Action: Laterally rotates the thigh at the hip joint

Innervation: L5–S2

Palpation: Deep muscle; cannot be palpated.

This is a thin, strap-like muscle. **Synergists** are the **inferior gemellus, quadratus femoris,** and **internal** and **external obturators;** its **antagonist** is the **gluteus minimus.**

Gemellus Inferior
(jee•**mel**•lus)

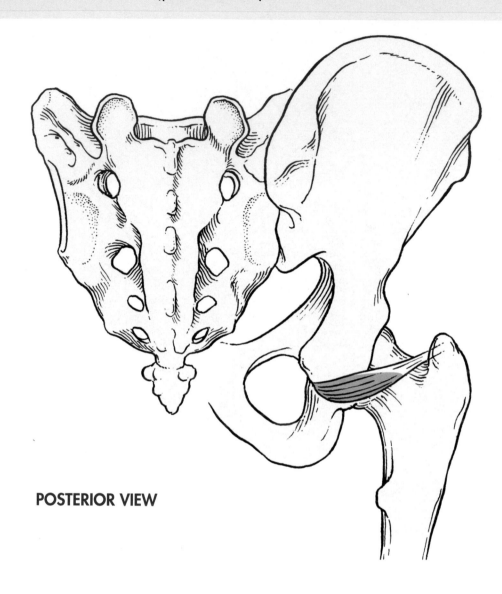

POSTERIOR VIEW

Origin (proximal attachment): Upper margin of ischial tuberosity

Insertion (distal attachment): With tendon of obturator internus into upper border of greater trochanter

Action: Laterally rotates thigh at hip joint

Innervation: Nerve from sacral plexus (L5–S2)

Palpation: Deep muscle; cannot be palpated.

Synergists are the **superior gemellus, quadratus femoris, internal** and **external obturators,** and **piriformis;** its **antagonist** is the **gluteus minimus.**

Quadratus Femoris
(kwa•**drah**•tus) (**fe**•moh•ris)

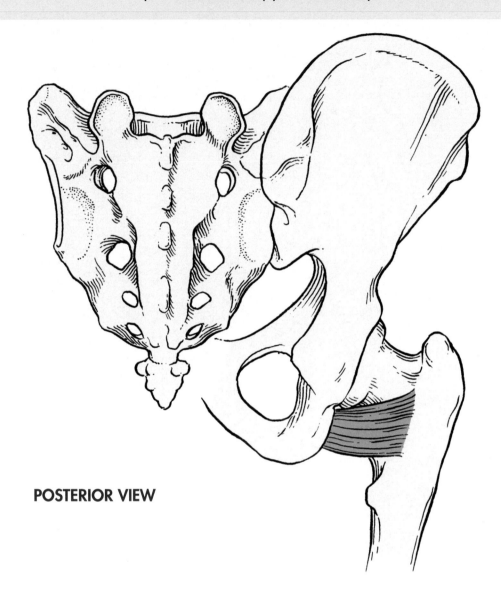

POSTERIOR VIEW

Origin (proximal attachment): Upper part of the lateral border of the ischial tuberosity

Insertion (distal attachment): Trochanteric crest of femur

Action: Laterally rotates the thigh at the hip joint

Innervation: Branch from sacral plexus (L5, S1)

Palpation: Palpate between ischial tuberosity and intertrochanteric crest of femur during active lateral rotation of thigh.

This is the most inferior of the rotators of the hip. **Synergists** are the **superior** and **inferior gemelli, piriformis, and internal** and **external obturators**; its **antagonist** is the **gluteus minimus.**

Gluteus Maximus
(gloo•te•us) (max•ih•mus)

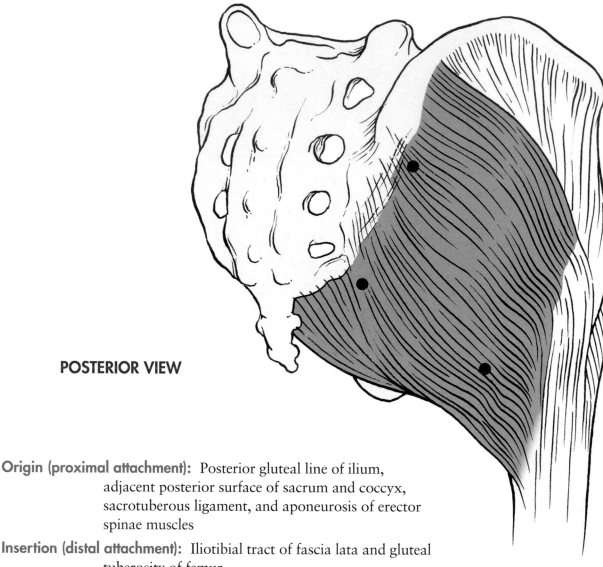

POSTERIOR VIEW

Origin (proximal attachment): Posterior gluteal line of ilium, adjacent posterior surface of sacrum and coccyx, sacrotuberous ligament, and aponeurosis of erector spinae muscles

Insertion (distal attachment): Iliotibial tract of fascia lata and gluteal tuberosity of femur

Action: **Upper part**—extends and laterally rotates thigh

Lower part—extends, laterally rotates thigh and assists in raising the trunk from a flexed position; also assists in adduction of the hip joint

Innervation: Inferior gluteal nerve (L5–S2)

Palpation: Palpate buttock during active extension and lateral rotation of thigh.

The **gluteus maximus** muscles are important in maintaining the upright posture. It is active primarily during strenuous activities such as running, jumping, and climbing. It has three main **trigger points:** one near the sacrum, one near the ischial tuberosity, and one in the belly of the muscle near the lower fibers. Its **referred pain pattern** is the entire gluteal region. The **gluteus maximus** is an important injection site. The sciatic nerve runs deep through it. **Synergists** are the **hamstring muscles; antagonists** are the **adductor group,** and **rectus femoris.**

Gluteus Medius
(gloo•te•us) (me•dee•us)

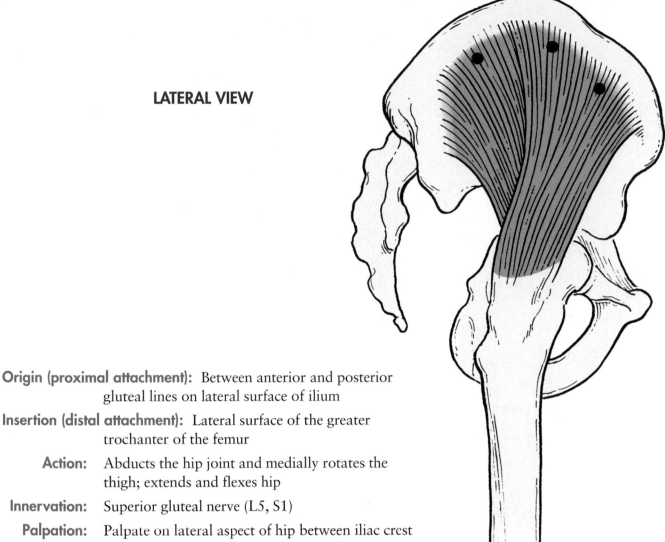

LATERAL VIEW

Origin (proximal attachment): Between anterior and posterior gluteal lines on lateral surface of ilium

Insertion (distal attachment): Lateral surface of the greater trochanter of the femur

Action: Abducts the hip joint and medially rotates the thigh; extends and flexes hip

Innervation: Superior gluteal nerve (L5, S1)

Palpation: Palpate on lateral aspect of hip between iliac crest and greater trochanter during active abduction.

The gluteal region is a common site for intramuscular injections because the muscles are thick and large. To avoid the sciatic nerve, gluteal nerves, and blood vessels deep in the **gluteus maximus,** injections are applied to the **gluteus medius** in the superolateral part of the buttock where the g. medius is not covered by the g. maximus. The **trigger points** are along the musculotendinous junction at the iliac crest. Its **referred pain pattern** is to the lower back and posterior and lateral areas of the buttock. **Synergists** are the **gluteus minimus** and **tensor fasciae latae; antagonists** are the **adductor group, gracilis,** and **pectineus.**

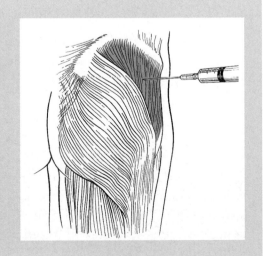

Gluteus Minimus
(gloo•te•us) (min•ih•mus)

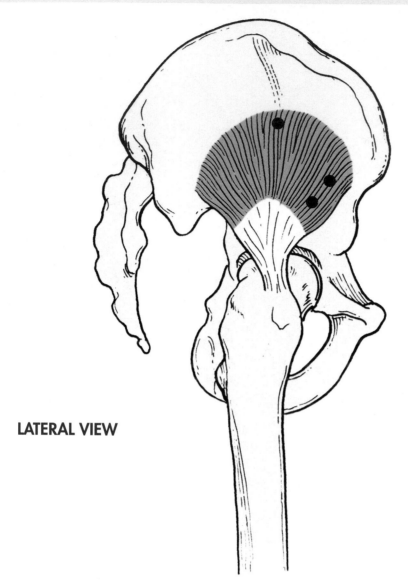

LATERAL VIEW

Origin (proximal attachment): Outer surface of the ilium between the middle and inferior gluteal lines

Insertion (distal attachment): Anterior border of the greater trochanter

Action: Abducts the femur at the hip joint and medially rotates the thigh; flexes hip

Innervation: Superior gluteal nerve (L4–S1)

Palpation: Deep muscle; cannot readily be palpated.

This is the smallest and deepest of the gluteal muscles. The two muscles together prevent the pelvis from dropping toward the opposite side during walking. It also keeps the pelvis level when standing on one foot. Its **trigger point** is in the belly of the muscle. Its **referred pain pattern** is the lower lateral buttock down the lateral aspect of the thigh, lower leg to the ankle. **Synergists** are the **gluteus medius** and **tensor fasciae latae**; **antagonists** are the **adductor group, pectineus,** and **gracilis.**

Tensor Fasciae Latae
(ten•ser) (fash•ee•a) (lay•tee)

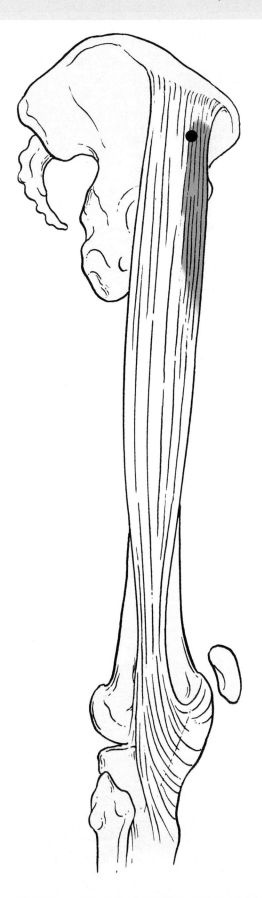

LATERAL VIEW

Origin (proximal attachment): Anterior aspect of the outer lip of the iliac crest and the anterior superior iliac spine

Insertion (distal attachment): Middle and proximal thirds of the thigh along the iliotibial tract. The iliotibial band inserts on the lateral epicondyle of tibia.

Action: Assists in abduction, medial rotation, and flexion of thigh. Makes the iliotibial tract taut. Stabilizer of the hip.

Innervation: Superior gluteal nerve (L4–S1)

Palpation: Palpate below superior anterior iliac spine at level of greater trochanter during hip abduction and flexion.

This muscle braces the knee when walking. Its **trigger point** is in the belly of the muscle near its proximal attachment. The **referred pain pattern** is localized in the hip and down the lateral side of the leg to the knee. It is one of several muscles (iliacus, rectus femoris, sartorius) implicated in a **"high pointer"** injury which may involve both a bruise on the bone of the iliac crest as well as tearing or avulsion of the attachments of these muscles to the iliac crest. It is **synergistic** with the **gluteus medius** and **minimus** for **abduction** of the thigh and with the **gluteus maximus** for **flexion** at the hip; **antagonists** are the **adductor group**, **gracilis**, and **pectineus**.

Sartorius

(sar•**tor**•ee•us)

This is a strap-like muscle running obliquely across the anterior surface of the thigh to the knee. It is the longest in the body. It crosses both the hip and knee joint. It is sometimes called the **tailor's muscle** and is used in sitting on the floor with thighs spread and lower legs crossed similar to a yoga position. Its **trigger points** are in three or four places in the long belly of the muscle. Its **referred pain pattern** is the entire anterior thigh with concentration at the knee. **Synergists** are the **hamstring muscles** and **gracilis**; **antagonists** are the **gluteal muscles, adductor muscles**, and the **tensor fascia latae**. It mainly assists the other muscles in their movement. Rarely, it is divided into two, inferiorly or completely along its length. The accessory portion may arise from various points near the anterior superior iliac spine, the pectineal line, pubic bone, or inguinal ligament. Insertions vary between adjacent areas of the femur, patellar ligament, tendon of the semitendinosus, or tendon of the twin muscle.

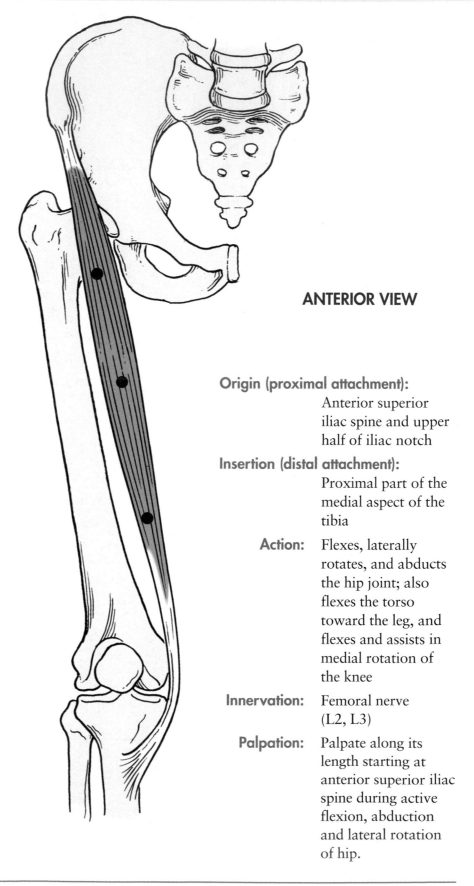

ANTERIOR VIEW

Origin (proximal attachment):
Anterior superior iliac spine and upper half of iliac notch

Insertion (distal attachment):
Proximal part of the medial aspect of the tibia

Action:
Flexes, laterally rotates, and abducts the hip joint; also flexes the torso toward the leg, and flexes and assists in medial rotation of the knee

Innervation:
Femoral nerve (L2, L3)

Palpation:
Palpate along its length starting at anterior superior iliac spine during active flexion, abduction and lateral rotation of hip.

Rectus Femoris
(**rek**•tus) (**fe**•moh•ris)

This is one of the four **quadriceps femoris** muscles. It sits on the anterior aspect of the femur and is the only one of the four that crosses both the hip and knee joint. This group is a powerful knee extensor used in running, jumping, climbing, and rising from a sitting position. Its **trigger points** are near its attachments. Its **referred pain pattern** is the entire anterior thigh with a concentration at the knee. The rectus femoris is used when thigh flexion and leg extension are needed such as kicking a soccer ball or football. The pain and muscle stiffness associated with a "**charley horse**" results from a contusion and tearing of muscle tissue of a quadriceps muscle such as the rectus femoris. Direct trauma of a hockey stick or tackle may produce a thigh hematoma, an accumulation of blood into the muscle and surrounding tissues from damaged blood vessels. **Synergists** are the other three **quadriceps femoris** muscles, psoas, and **sartorius**; **antagonists** are the **hamstring muscles.**

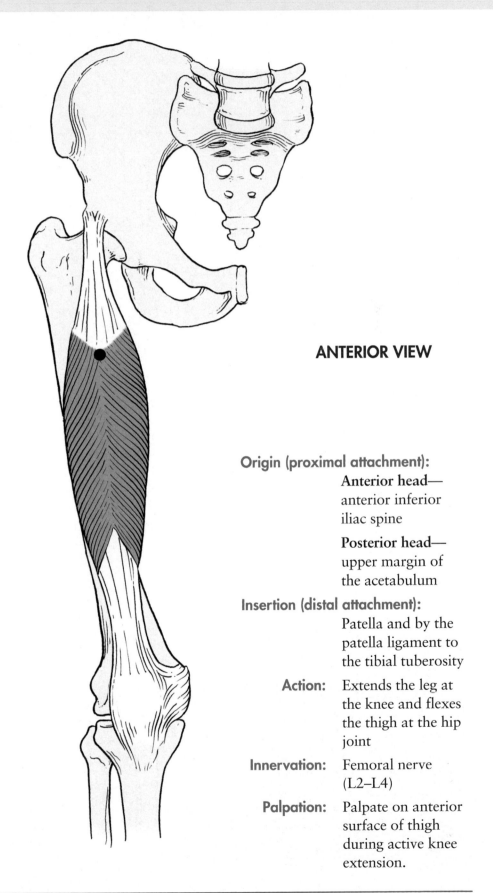

ANTERIOR VIEW

Origin (proximal attachment):
> **Anterior head**— anterior inferior iliac spine

> **Posterior head**— upper margin of the acetabulum

Insertion (distal attachment):
> Patella and by the patella ligament to the tibial tuberosity

Action: Extends the leg at the knee and flexes the thigh at the hip joint

Innervation: Femoral nerve (L2–L4)

Palpation: Palpate on anterior surface of thigh during active knee extension.

Vastus Medialis
(vas•tus)(mee•dee•ah•lis)

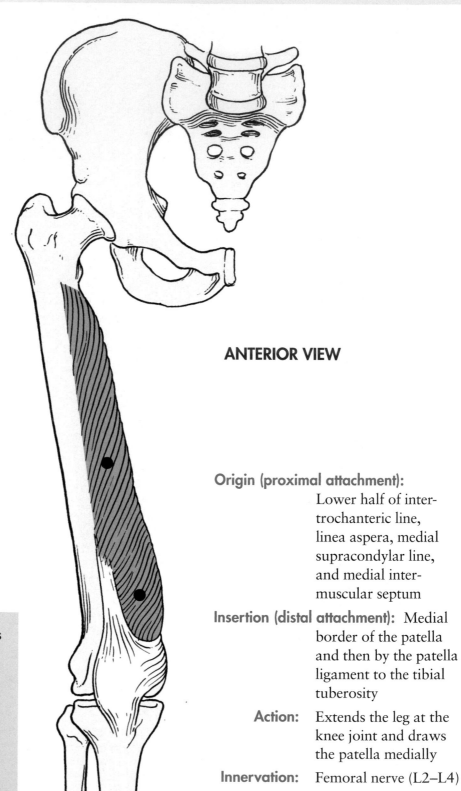

ANTERIOR VIEW

This is the most medial of the muscles of the **quadriceps femoris** group. It forms the medial aspect of the thigh. Its **trigger points** are in the belly and just above the insertion. Its **referred pain pattern** is the entire anterior thigh, especially the lower medial aspect with the most concentrated pain in the knee region. **Synergists** are the other three **quadriceps femoris muscles** and **sartorius**; antagonists are the **hamstring muscles**.

Origin (proximal attachment): Lower half of inter-trochanteric line, linea aspera, medial supracondylar line, and medial inter-muscular septum

Insertion (distal attachment): Medial border of the patella and then by the patella ligament to the tibial tuberosity

Action: Extends the leg at the knee joint and draws the patella medially

Innervation: Femoral nerve (L2–L4)

Palpation: Palpate on anterio-medial surface of lower third of thigh during active knee extension.

Vastus Lateralis
(vas•tus) (lat•er•ah•lis)

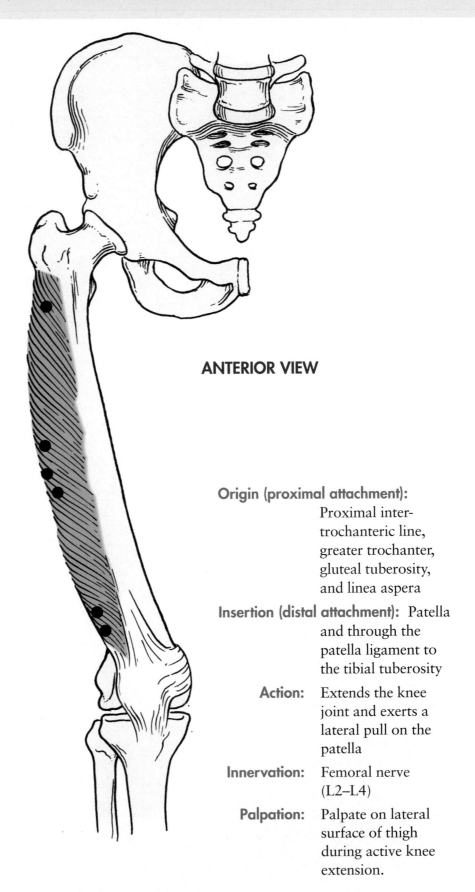

ANTERIOR VIEW

This is the most lateral of the **quadriceps femoris** group. It forms the lateral aspect of the thigh and is a common site for intramuscular injections. Striking the patella ligament causes the characteristic knee jerk reflex test. It has **trigger points** near each attachment and in the belly of the muscle. Its **referred pain pattern** is the anterior thigh especially the lateral surface and again with pain concentrated in the knee. Compression of the knee, or quadriceps muscle imbalance, may pull the patella sideways and produce the condition **chondromalacia patellae** or "**runners knee.**" It is a common condition in runners and may also be caused by a blow to the patella or extreme flexing of the knee joint as in squatting or kneeling. **Synergists** are the three **quadriceps femoris muscles** and **sartorius; antagonists** are the **hamstring muscles.**

Origin (proximal attachment): Proximal inter-trochanteric line, greater trochanter, gluteal tuberosity, and linea aspera

Insertion (distal attachment): Patella and through the patella ligament to the tibial tuberosity

Action: Extends the knee joint and exerts a lateral pull on the patella

Innervation: Femoral nerve (L2–L4)

Palpation: Palpate on lateral surface of thigh during active knee extension.

Vastus Intermedius
(vas•tus) (in•ter•mee•de•us)

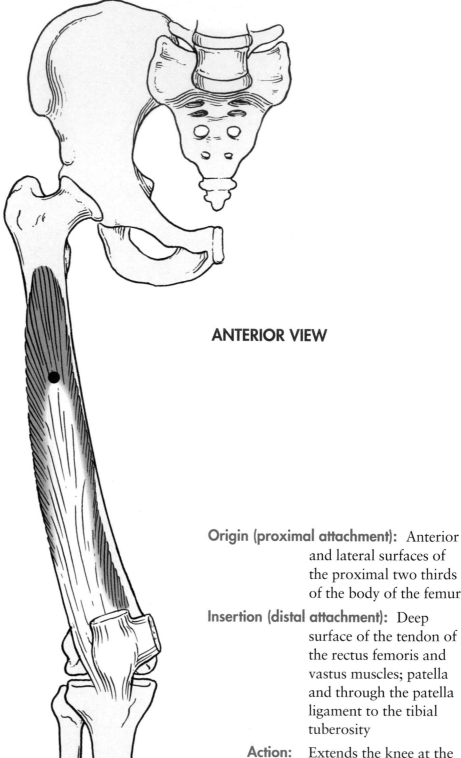

ANTERIOR VIEW

This muscle is the smallest and deepest of the **quadriceps femoris** group. It is covered by the **rectus femoris** and lies between the two vastus muscles. Its **trigger point** is near the proximal attachment. Its **referred pain pattern** is the deep anterior thigh. **Synergists** are the other three **quadriceps femoris muscles** and **sartorius; antagonists** are the **hamstring muscles.**

Origin (proximal attachment): Anterior and lateral surfaces of the proximal two thirds of the body of the femur

Insertion (distal attachment): Deep surface of the tendon of the rectus femoris and vastus muscles; patella and through the patella ligament to the tibial tuberosity

Action: Extends the knee at the joint

Innervation: Femoral nerve (L2–L4)

Palpation: Deep muscle; cannot readily be palpated.

An Illustrated Atlas of the Skeletal Muscles

Biceps Femoris
(**bi**•seps) (**fe**•moh•ris)

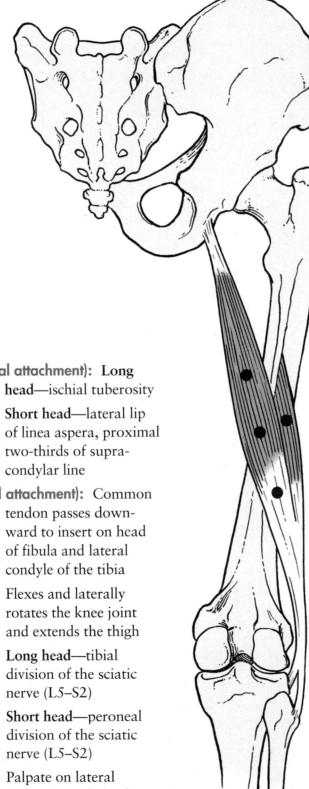

POSTERIOR VIEW

Origin (proximal attachment): **Long head**—ischial tuberosity

Short head—lateral lip of linea aspera, proximal two-thirds of supra-condylar line

Insertion (distal attachment): Common tendon passes downward to insert on head of fibula and lateral condyle of the tibia

Action: Flexes and laterally rotates the knee joint and extends the thigh

Innervation: **Long head**—tibial division of the sciatic nerve (L5–S2)

Short head—peroneal division of the sciatic nerve (L5–S2)

Palpation: Palpate on lateral posterior surface of thigh during resisted knee flexion.

This muscle is the most lateral of the **hamstring** group. The two heads may be more separated lengthwise than normal or the short head may be absent. It crosses both the hip and knee joints, and it is a prime mover of both hip extension and knee flexion. "**Pulled hamstrings**" is a common sports injury. It occurs with hard running, fast starts, or kicking. A contusion occurs with tearing of the muscle fibers and a hematoma forming. The hematoma is contained by the dense fascia of the muscle. Occasionally the tears are in the tendons attaching it to the ischial tuberosity. Its **trigger points** are in the belly of the muscle and near the insertion. Its **referred pain pattern** is from the ischial tuberosity to the back of the knee and down the posterior leg to mid-calf. **Synergists** are the other **hamstring muscles, gracilis, gastrocnemius,** and **sartorius; antagonists** are the four **quadriceps femoris muscles.**

Semitendinosus
(sem•ee•ten•dih•no•sus)

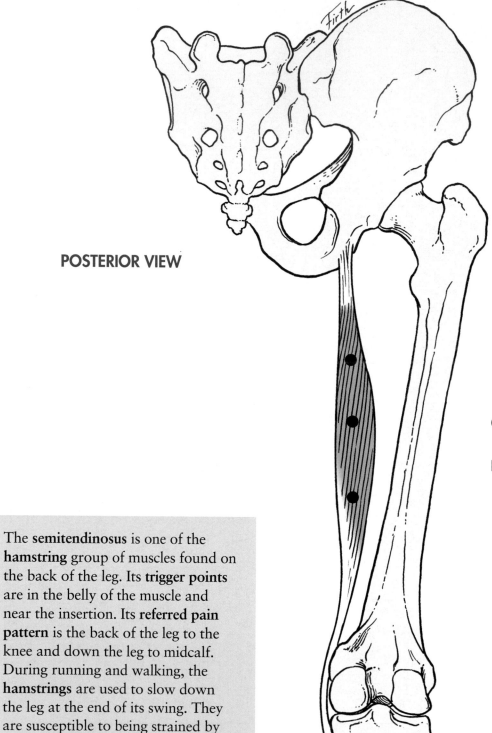

POSTERIOR VIEW

Firth

The **semitendinosus** is one of the **hamstring** group of muscles found on the back of the leg. Its **trigger points** are in the belly of the muscle and near the insertion. Its **referred pain pattern** is the back of the leg to the knee and down the leg to midcalf. During running and walking, the **hamstrings** are used to slow down the leg at the end of its swing. They are susceptible to being strained by resisting the momentum of the action. **Synergists** are the other **hamstring muscles, gracilis, gastrocnemius,** and **sartorius; antagonists** are the four **quadriceps femoris muscles.**

Origin (proximal attachment): Ischial tuberosity

Insertion (distal attachment): Upper medial surface of the shaft of the tibia

Action: Flexes and slightly medially rotates leg at knee joint, and extends the thigh at the hip joint

Innervation: Tibial portion of sciatic nerve (L5–S2)

Palpation: Palpate on posterior lower thigh down to medial aspect of popliteal fossa during active knee flexion.

Semimembranosus

(**sem**•ee•**mem**•brah•**no**•sus)

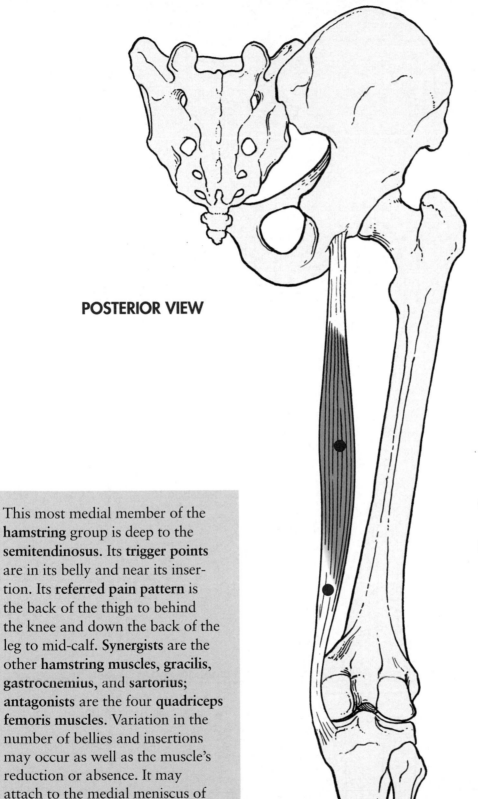

POSTERIOR VIEW

This most medial member of the **hamstring** group is deep to the **semitendinosus**. Its **trigger points** are in its belly and near its insertion. Its **referred pain pattern** is the back of the thigh to behind the knee and down the back of the leg to mid-calf. **Synergists** are the other **hamstring muscles, gracilis, gastrocnemius,** and **sartorius; antagonists** are the four **quadriceps femoris muscles.** Variation in the number of bellies and insertions may occur as well as the muscle's reduction or absence. It may attach to the medial meniscus of the knee, facilitating movement of the meniscus during flexion.

Origin (proximal attachment):
    Ischial tuberosity

Insertion (distal attachment):
    Posterior part of the medial condyle of tibia

Action: Flexes and slightly medially rotates leg at knee joint and extends thigh at hip

Innervation: Tibial portion of sciatic nerve (L5–S2)

Palpation: Palpate on posterior medial thigh during active knee flexion.

Gracilis
(grah•sil•is)

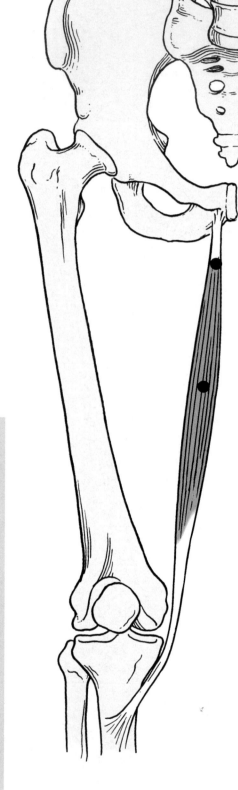

ANTERIOR VIEW

The **gracilis** is a long slender superficial muscle of the medial thigh. Its **trigger points** are found in the belly of the muscle and near the origin. Its **referred pain pattern** is deep into the groin, into the medial thigh, and downward to the knee and shin. Because it is a relatively weak member of the thigh adductor group, the gracilis, together with its nerves and blood vessels, has been transplanted to replace a damaged muscle. **Synergists** are the **adductor muscles; antagonists** are the **tensor fasciae latae** and **gluteal muscles.**

Origin (proximal attachment): Inferior ramus and body of pubis

Insertion (distal attachment): Medial surface of tibia just inferior to its medial condyle

Action: Adducts thigh at hip joint and flexes leg at knee joint; assists in medial rotation and flexes hip

Innervation: Obturator nerve (L3, L4)

Palpation: Palpate muscle on the upper medial side of the thigh during active hip adduction; palpate its tendon on medial side of knee medial to tendon of the semitendinosus.

Pectineus
(pek•tin•ee•us)

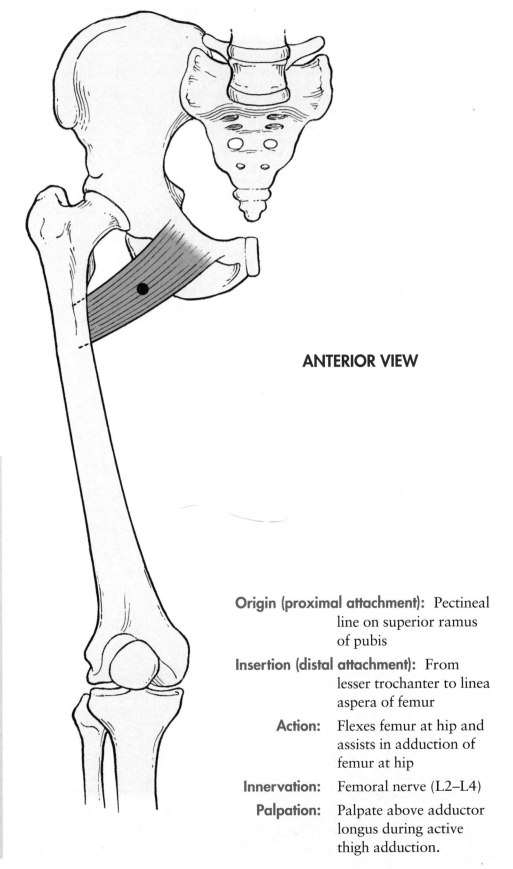

ANTERIOR VIEW

The **pectineus** is the uppermost of the **adductor** group of muscles. There is controversy about whether it medially rotates the thigh. Straining of this muscle causes a **"pulled groin"** injury. All of the **adductor** group are important in horseback riding or other activities that require the pressing together of the thighs. The **trigger point** is in the belly of the muscle. Its **referred pain pattern** is deep in the groin area. **Synergists** are the other **adductor muscles** and **gracilis; antagonists** are the **tensor fasciae latae** and **gluteal minimus** and **medius muscles.**

Origin (proximal attachment): Pectineal line on superior ramus of pubis

Insertion (distal attachment): From lesser trochanter to linea aspera of femur

Action: Flexes femur at hip and assists in adduction of femur at hip

Innervation: Femoral nerve (L2–L4)

Palpation: Palpate above adductor longus during active thigh adduction.

Adductor Brevis
(ad•**duck**•ter) (**brev**•is)

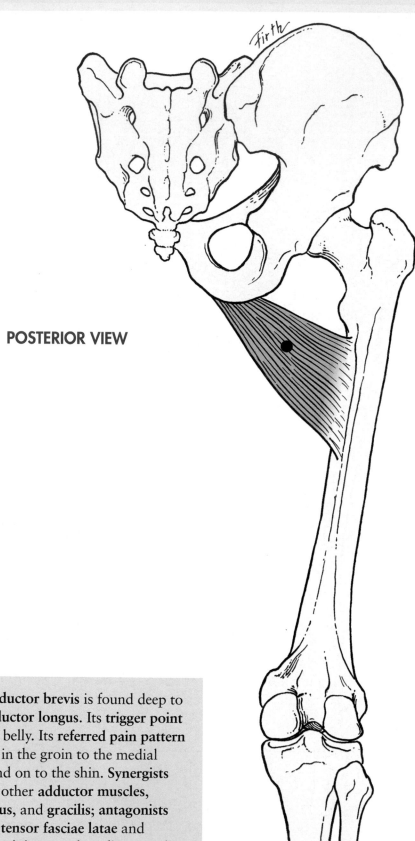

POSTERIOR VIEW

The **adductor brevis** is found deep to the **adductor longus**. Its **trigger point** is in its belly. Its **referred pain pattern** is deep in the groin to the medial knee and on to the shin. **Synergists** are the other **adductor muscles**, **pectineus**, and **gracilis**; antagonists are the **tensor fasciae latae** and **gluteal minimus** and **medius** muscles.

Origin (proximal attachment):
Outer surface of inferior ramus of pubis

Insertion (distal attachment):
Upper one third of medial lip of the linea aspera of the femur

Action: Adducts the thigh. Assists in flexion and medial rotation.

Innervation: Obturator nerve (L3, L4)

Palpation: Deep muscle; cannot be readily palpated.

Adductor Longus
(ad•**duck**•ter) (**lon**•gus)

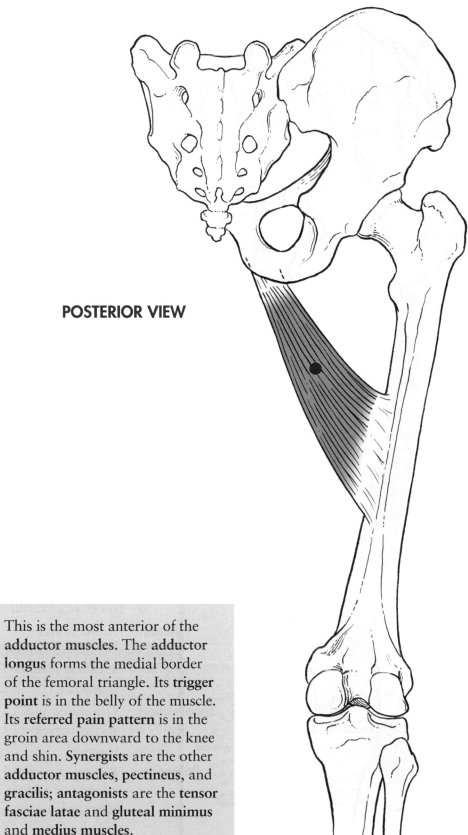

POSTERIOR VIEW

This is the most anterior of the **adductor muscles**. The **adductor longus** forms the medial border of the femoral triangle. Its **trigger point** is in the belly of the muscle. Its **referred pain pattern** is in the groin area downward to the knee and shin. **Synergists** are the other **adductor muscles, pectineus,** and **gracilis; antagonists** are the **tensor fasciae latae** and **gluteal minimus** and **medius muscles.**

Origin (proximal attachment):
Anterior body of pubis

Insertion (distal attachment):
Medial one third of medial lip of linea aspera of femur

Action: Adducts and flexes thigh; assists in medial rotation

Innervation: Obturator nerve (L3, L4)

Palpation: Palpate in medial aspect of groin during active adduction of femur at hip.

Tibialis Posterior
(tib•ee•a•lis)

The **trigger point** is in the belly of the muscle near the knee. Its **referred pain pattern** is in the knee and down the posterior leg. **Synergists** are the **flexor hallucis longus** and **flexor digitorum longus; antagonists** are the **extensor hallucis longus, extensor digitorum longus,** and **tibialis anterior.**

Origin (proximal attachment): Lateral part of posterior surface of tibia, interosseous membrane, and proximal half of posterior surface of fibula

Insertion (distal attachment): Tuberosity of navicular bone, cuboid, cuneiforms, second, third, and fourth metatarsals and calcaneus

Action: Plantar flexion and inversion of the foot

Innervation: Tibial nerve (L5, S1)

Palpation: Palpate tendon on medial malleolus during active inversion of foot.

POSTERIOR VIEW

An Illustrated Atlas of the Skeletal Muscles

Tibialis Anterior
(tib•ee•a•lis)

This is a superficial muscle of the shin. It parallels laterally the sharp anterior margin of the tibia. Paralysis of this muscle causes **"foot drop"** in which the foot must be lifted high to prevent the toes from dragging during walking. Irritation of this muscle during running on hard surfaces often causes **"shin splints."** Shin splints occur in the anterior compartment of the leg following vigorous exercise in someone who does not usually exercise. The tibialis anterior muscle swells and the blood supply is reduced causing ischemia. Proper warm up and cool down techniques help prevent this problem. The **trigger point** for this muscle is in its belly. The **referred pain pattern** is down the shin to the ankle and into the toes. **Synergists** are the **extensor hallucis longus** and **extensor digitorum longus**; **antagonists** are the **gastrocnemius, soleus,** and **fibularis longus.**

ANTERIOR VIEW

Origin (proximal attachment): Lateral condyle and proximal one half of the lateral surface of the tibia and the interosseous membrane

Insertion (distal attachment): Medial plantar surface of the first cuneiform bone and the base of the first metatarsal bone

Action: Dorsiflexion of the foot at the ankle joint and inversion of the foot

Innervation: Deep peroneal nerve (L4–S1)

Palpation: Palpate muscle on anterior surface of tibia during active dorsiflexion of foot; palpate its tendon on medial side of anterior surface of ankle.

Extensor Digitorum Longus

(ex•**ten**•ser) (dij•ih•**tor**•um) (**lon**•gus)

Origin (proximal attachment): Lateral condyle of the tibia, proximal three fourths of the anterior surface of the fibula, and the interosseous membrane

Insertion (distal attachment): By four tendons to the second through fifth digits. Each tendon divides into a middle slip that inserts in the base of the middle phalanx and two lateral slips that insert in the base of the distal phalanx

Action: Extends the phalanges of the second through fifth digits, assists in dorsiflexion of the ankle and in eversion of the foot

Innervation: Deep peroneal nerve (L4–S1)

Palpation: Palpate common tendon on anterior surface of ankle lateral to extensor hallucis longus tendon during active extension of toes; palpate divided tendon on dorsal surface of foot.

The **trigger point** is located near the origin of the muscle, with the **referred pain distribution** mainly to the top of the foot. **Synergists** are the **extensor digitorum brevis** and **tibialis anterior**; antagonists are the **flexor digitorum longus, gastrocnemius,** and **soleus.**

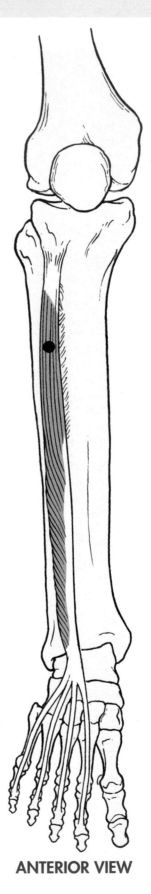

ANTERIOR VIEW

Extensor Hallucis Longus
(ex•ten•ser) (hal•lu•sis) (lon•gus)

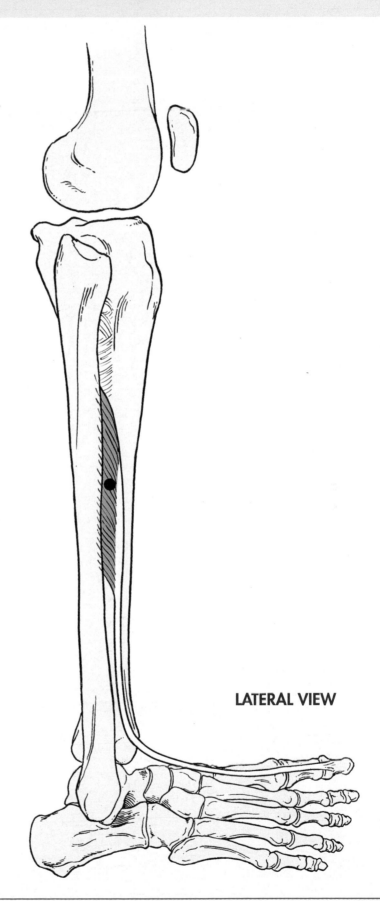

LATERAL VIEW

Origin (proximal attachment): Middle half of anterior surface of fibula and adjacent interosseous membrane

Insertion (distal attachment): Base of the distal phalanx of the big toe

Action: Extends or dorsiflexes the big toe; it aids in dorsi-flexing the foot.

Innervation: Deep peroneal nerve (L4–S1)

Palpation: Palpate tendon lateral to tibialis anterior tendon on anterior surface of ankle and also on dorsum of foot near big toe.

The **trigger point** is in the belly. Its **referred pain pattern** is down the side of the leg into the big toe. Its **synergist** is the **extensor hallucis brevis**; its **antagonist** is the **flexor hallucis longus**.

Peroneus Longus—Fibularis Longus
(perr•o•nee•us) (lon•gus) (fib•yoo•lair•iss) (lon•gus)

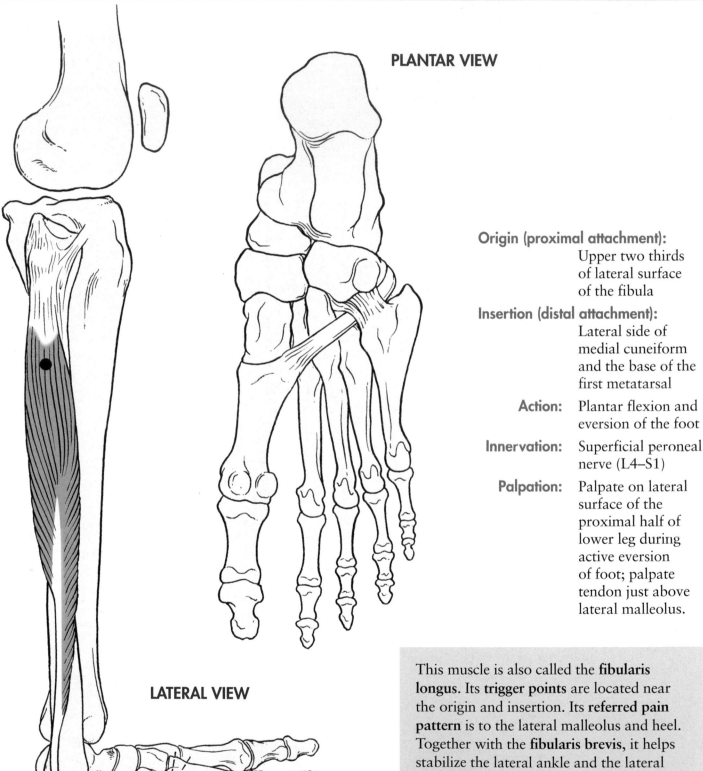

PLANTAR VIEW

LATERAL VIEW

Origin (proximal attachment): Upper two thirds of lateral surface of the fibula

Insertion (distal attachment): Lateral side of medial cuneiform and the base of the first metatarsal

Action: Plantar flexion and eversion of the foot

Innervation: Superficial peroneal nerve (L4–S1)

Palpation: Palpate on lateral surface of the proximal half of lower leg during active eversion of foot; palpate tendon just above lateral malleolus.

This muscle is also called the **fibularis longus**. Its **trigger points** are located near the origin and insertion. Its **referred pain pattern** is to the lateral malleolus and heel. Together with the **fibularis brevis**, it helps stabilize the lateral ankle and the lateral longitudinal arch of the foot. **Synergists** for **plantar flexion** are the **gastrocnemius** and **soleus** and for eversion are the **fibularis brevis** and **tertius**.

Peroneus Brevis—Fibularis Brevis

(perr•o•**nee**•us)(**brev**•is)   (**fib**•yoo•lair•is) (**brev**•is)

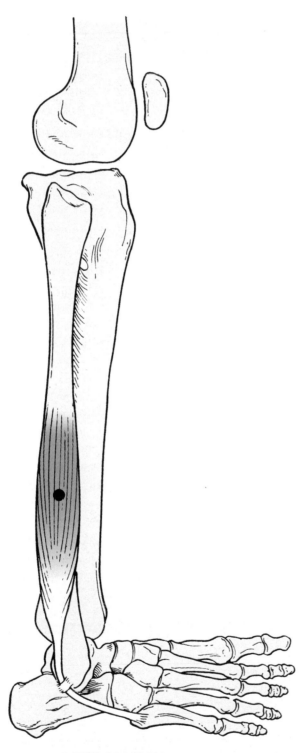

LATERAL VIEW

Origin (proximal attachment): Lower two thirds of lateral surface of the fibula

Insertion (distal attachment): Lateral side of the base of the fifth metatarsal bone

Action: Plantar flexion and eversion of the foot

Innervation: Superficial peroneal nerve (L4–S1)

Palpation: Palpate tendon on lateral dorsal surface of foot near proximal end of fifth metatarsal.

This muscle is also called the **fibularis brevis**. The action of the foot "evertors" is helpful when walking or running on uneven surfaces. Its **trigger point** is in the belly of the muscle. Its **referred pain pattern** is the lateral malleolus. **Synergists** for **plantar flexion** are the **gastrocnemius** and **soleus** and for **eversion** are the **fibularis longus** and **tertius**.

Peroneus Tertius—Fibularis Tertius

(perr•o•**nee**•us)(**ter**•she•us)(**fib**•yoo•lair•is) (**ter**•she•us)

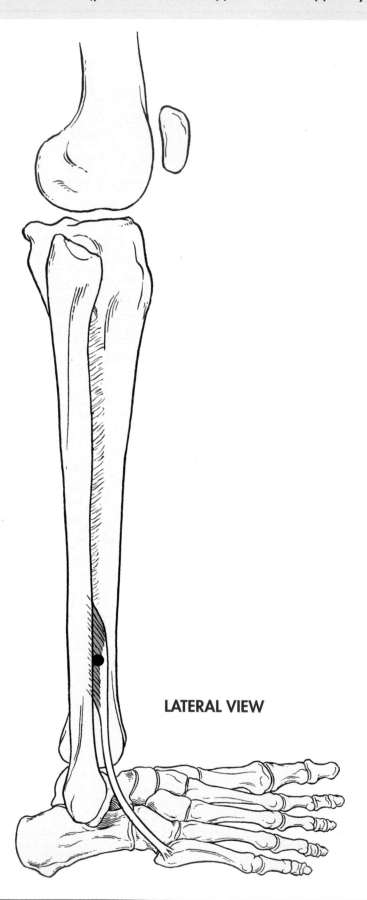

LATERAL VIEW

Origin (proximal attachment): Lower third of anterior surface of the fibula and the interosseous membrane

Insertion (distal attachment): Dorsal surface of the base of the fifth metatarsal bone

Action: Dorsiflexion and eversion of the foot

Innervation: Deep peroneal nerve (L4–S1)

Palpation: Palpate tendon lateral to the tendon of the extensor digitorum longus on dorsal surface of foot at base of fifth metatarsal bone.

This muscle may also insert onto the base of the fourth metatarsal. The **trigger point** is located in the belly of the muscle. Its **synergist** for **dorsiflexion** is the **extensor digitorum longus** and for **eversion** is the **fibularis longus** and **brevis**; its **antagonist** for **dorsiflexion** are the **gastrocnemius** and **soleus** and for **eversion** is the **tibialis anterior**.

Extensor Digitorum Brevis and Extensor Hallucis Brevis

(ex•**ten**•ser) (dij•ih•**tor**•um) (**brev**•is)   (ex•**ten**•ser) (hal•**lu**•sis) (**brev**•is)

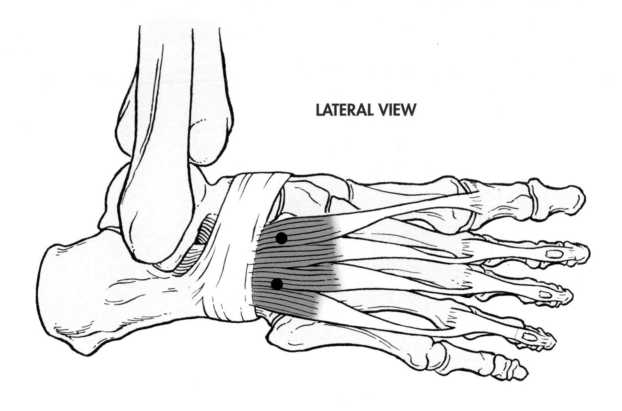

LATERAL VIEW

Origin (proximal attachment): Anterior and lateral surfaces of the calcaneous, lateral talocalcaneal ligament, and inferior extensor retinaculum

Insertion (proximal attachment): Dorsal surface of the base of the proximal phalanx of the big toe and the lateral sides of the tendons of the extensor digitorum longus of the second, third, and fourth toes

Action: Extends the metatarsalphalangeal joint of the big toe and extends the interphalangeal and metatarsophalangeal joints of the second through fourth toes.

Innervation: Deep peroneal nerve (L5–S2)

Palpation: Palpate distal and anterior to medial malleolus during extension of toes.

The **trigger points** are located toward the origin end of these short toe extensors. The **referred pain pattern** occurs right around these muscles on the outer side of the top of the foot. **Synergists** are the **extensor digitorum longus** and **extensor hallucis longus**; antagonists are the **flexor digitorum muscles** and **flexor hallucis muscles.**

Flexor Digitorum Brevis
(**flex**•er) (dij•ih•**tor**•um) (**brev**•is)

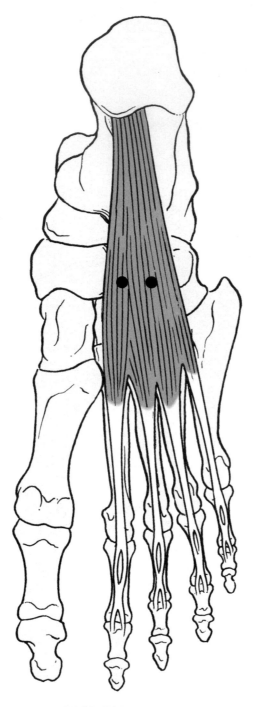

PLANTAR VIEW

Origin (proximal attachment): Tuberosity of the calcaneous and plantar apononeurosis

Insertion (distal attachment): Sides of middle phalanges of the second through fifth toes

Action: Flexion of proximal phalanges

Innervation: Medial plantar nerve (L4, L5)

Palpation: Palpate medially on plantar surface of foot when toes are flexed.

This is one of the superficial muscles of the sole of the foot. Together these muscles help support the arches of the foot. The slip to the little toe may be reduced or absent. The **trigger points** are in the belly of the muscle as it divides into slips onto the second through fifth toes. Its **synergist** is the **flexor digitorum longus**; its **antagonist** is the **extensor digitorum longus**.

Abductor Hallucis
(ab•**duck**•ter) (hal•**lu**•sis)

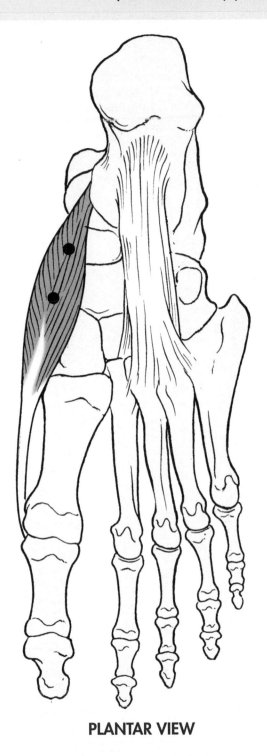

PLANTAR VIEW

Origin (proximal attachment): Tuberosity of calcaneous, flexor retinaculum, and plantar aponeurosis

Insertion (distal attachment): Medial side of base of proximal phalanx of big toe

Action: Abducts and assists in flexion at the metatarsophalangeal joint of the big toe

Innervation: Medial plantar nerve (L4, L5)

Palpation: Palpate along medial border of big toe when toe is actively abducted.

The muscles of the sole of the foot can be divided into four layers from most superficial to deepest. The most superficial layer includes this muscle and the **flexor digitorum brevis** and the **abductor digiti minimi**. Its **trigger points** are in the belly of the muscle. This muscle may have muscle or tendon "slips" to the base of the first phalanx of the second and third toes. **Synergists** are the **flexors hallucis longus** and **brevis**; **antagonists** are the **extensors hallucis longus** and **brevis**, and **adductor hallucis**.

Abductor Digiti Minimi
(ab•**duck**•ter)(**dij**•ih•tye)(**min**•i•mih)

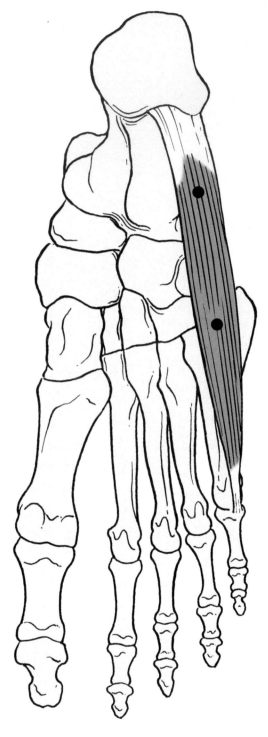

PLANTAR VIEW

Origin (proximal attachment): Tuberosity of the calcaneous and the plantar aponeurosis

Insertion (distal attachment): Lateral side of proximal phalanx of the fifth toe

Action: Abducts the fifth toe and flexes it at the metatarsophalangeal joint

Innervation: Lateral plantar nerve (S1, S2)

Palpation: Palpate along lateral border of little toe when toe is actively abducted.

This is one of the superficial muscles of the sole of the foot. The **trigger points** are near the origin and in the belly of the muscle. Its **synergist** is the **flexor digiti minimi; antagonists** are the **plantar interossei**.

Quadratus Plantae
(kwa•**drah**•tus) (**plan**•tay)

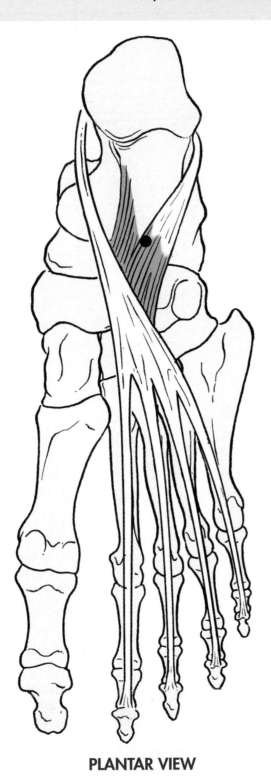

PLANTAR VIEW

Origin (proximal attachment): **Medial head**—medial surface of the calcaneous

Lateral head—lateral border of the inferior surface of the calcaneous

Insertion (distal attachment): Lateral margin of the tendon of the flexor digitorum longus

Action: Flexion of the terminal phalanges of the second through fifth toes

Innervation: Lateral plantar nerve (S1, S2)

Palpation: Deep muscle; cannot be palpated.

This is one of the second layer muscles of the sole of the foot. The **trigger point** is near the attachment to the calcaneous. **Synergists** are the **flexor digitorum longus** and **brevis**; antagonists are the **extensor digitorum longus** and **brevis**.

Lumbricales—Lumbricals
(lum•brih•kay•leez)

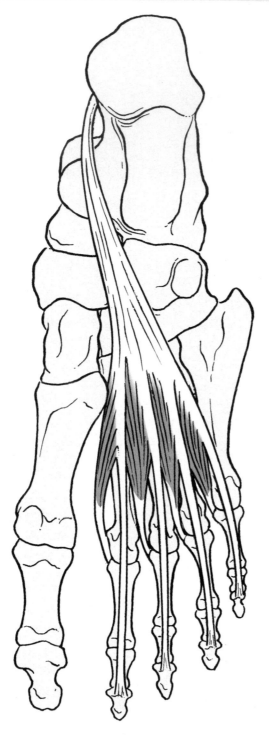

PLANTAR VIEW

Origin (proximal attachment): Tendons of the flexor digitorum longus

Insertion (distal attachment): Dorsal surfaces of the distal phalanges

Action: Flexion of second through fifth toes at the metatarsophalangeal joint

Innervation: **First lumbricalis—** medial plantar nerve (L4, L5)

Second through fifth lumbricales— lateral plantar nerve (S1, S2)

Palpation: Deep muscles; cannot be palpated.

This muscle is one of the second layer muscles in the sole of the foot. **Synergists** are the **flexors digitorum longus** and **brevis, quadratus plantae,** and **plantar interossei;** antagonists are the **extensors digitorum longus** and **brevis.**

Flexor Hallucis Brevis
(flex•er) (hal•lu•sis) (brev•is)

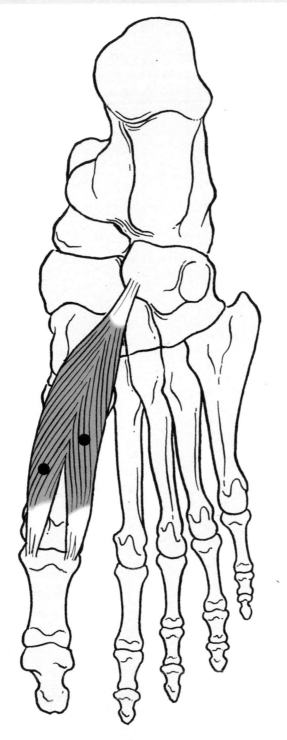

PLANTAR VIEW

Origin (proximal attachment):
 Plantar surface of cuboid and lateral cuneiform bones

Insertion (distal attachment): Medial part—medial side of the base of the proximal phalanx of the big toe

 Lateral part—lateral side of the base of the proximal phalanx of the big toe

Action: Flexion of the metatarsophalangeal joint of the big toe

Innervation: Medial plantar nerve (L4–S1)

Palpation: Deep muscle; cannot be palpated.

This muscle is one of the third layer muscles on the sole of the foot. The **trigger points** are located in the belly of each muscle slip. **Synergist** is the **flexor hallucis longus; antagonists** are the **extensor digitorum longus** and **brevis.**

Adductor Hallucis
(ad•**duck**•ter) (hal•**lu**•sis)

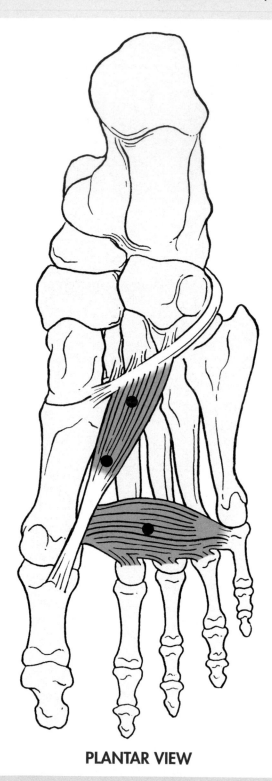

PLANTAR VIEW

Origin (proximal attachment):

> **Oblique head**—bases of second, third, and fourth metatarsal bones and sheath of fibularis longus tendon

> **Transverse head**—ligaments of the metatarsophalangeal joints of the third, fourth, and fifth toes

Insertion (distal attachment): Lateral side of base of proximal phalanx of the big toe

Action: Adducts the big toe

Innervation: Lateral plantar nerve (S1, S2)

Palpation: Deep muscle; cannot be palpated.

This muscle is one of the third layer muscles in the sole of the foot. It helps to maintain the transverse arch of the foot. The transverse head may also insert on the metatarsal of the big toe, allowing for opposition of the big toe. The transverse head may be absent. The **trigger points** are found in the belly of the transverse head, in the belly, and near the insertion of the oblique head. Its **antagonist** is the **abductor hallucis.**

An Illustrated Atlas of the Skeletal Muscles

Flexor Digiti Minimi Brevis
(flex•er)(dij•ih•tye)(min•ih•mih)(brev•is)

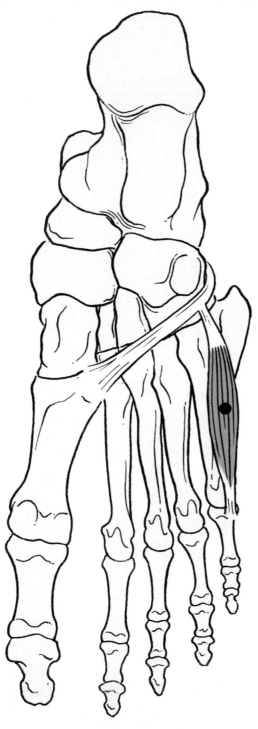

PLANTAR VIEW

Origin (proximal attachment): Base of the fifth metatarsal bone and sheath of the fibularis longus tendon

Insertion (distal attachment): Lateral side of base of the proximal phalanx of the fifth toe

Action: Flexion of the proximal phalanx of the fifth toe

Innervation: Lateral plantar nerve (S1, S2)

Palpation: Deep muscle; cannot be palpated.

Occasionally this muscle also arises from the cuboid, allowing opposition of the little toe. The **trigger point** is located in the belly of the muscle. **Synergists** are the **abductor digiti minimi** and **plantar interossei**; its **antagonist** is the **extensor digitorum longus**.

Plantar Interossei
(plan•tar)(in•ter•os•ee•eye)

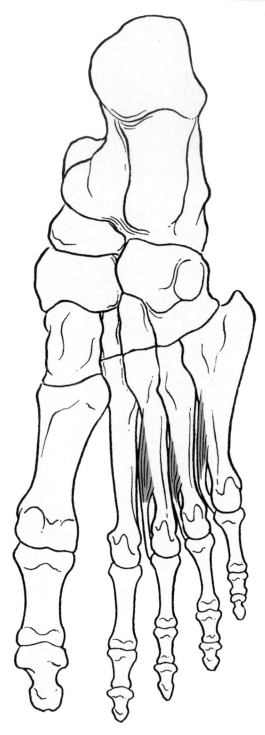

PLANTAR VIEW

Origin (proximal attachment): Bases and medial sides of the third, fourth, and fifth metatarsal bones

Insertion (distal attachment): Medial sides of bases of proximal phalanges

Action: Adducts toes and flexion of toes at metatarsalphalangeal joint

Innervation: Lateral plantar nerve (S1, S2)

Palpation: Deep muscles; can not be palpated.

Synergists for **flexion** are the **lumbricales, flexors digitorum longus** and **brevis,** and **quadratus plantae; antagonists** to **flexion** are the **extensors digitorum longus** and **brevis;** to **adduction** are the **dorsal interossei.**

An Illustrated Atlas of the Skeletal Muscles

Dorsal Interossei
(**dor**•sal)(**in**•ter•**os**•ee•eye)

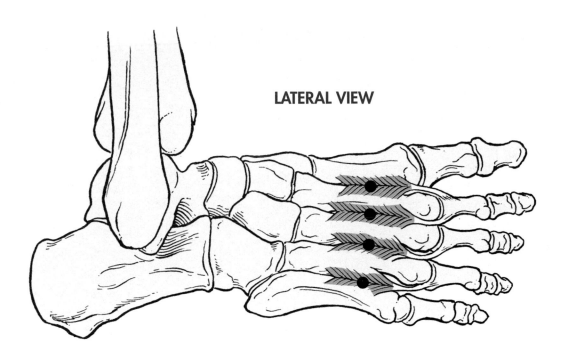

LATERAL VIEW

Origin (proximal attachment): Adjacent sides of metatarsal bones

Insertion (distal attachment): **First**—medial side of proximal phalanx of second toe

Second, third, and fourth—lateral sides of the bases of proximal phalanges of the second, third, and fourth toes

Action: Abduct toes and flexion of proximal phalanges at the metatarsophalangeal joint

Innervation: Lateral plantar nerve (S1, S2)

Palpation: Deep muscles; cannot be palpated.

The **trigger points** are in the belly of each muscle. **Synergists** for **flexion** are the **lumbricales, flexors digitorum longus** and **brevis,** and **quadratus plantae; antagonists** to **flexion** are the **extensors digitorum longus** and **brevis;** to **abduction** are the **plantar interossei.**

Functional Muscle Groups

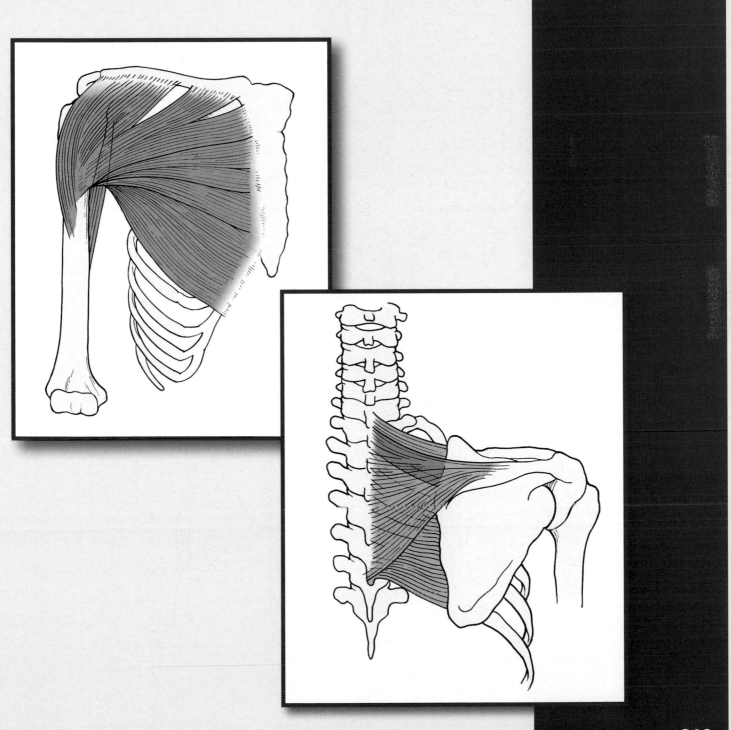

11

Elevators of Scapula
(ell•a•**vay**•tors) (**skap**•yoo•lah)

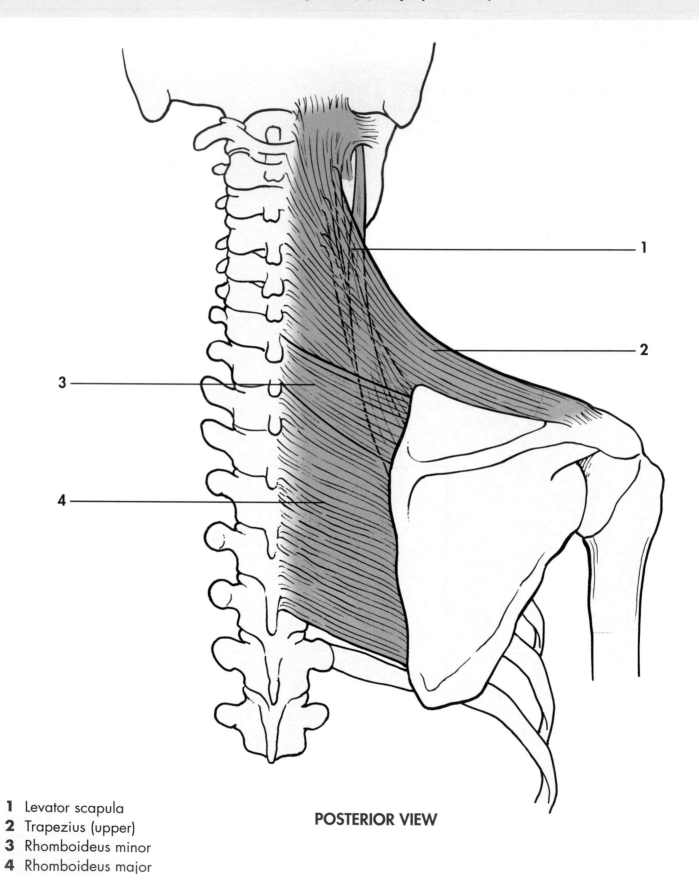

1

2

3

4

POSTERIOR VIEW

1 Levator scapula
2 Trapezius (upper)
3 Rhomboideus minor
4 Rhomboideus major

An Illustrated Atlas of the Skeletal Muscles

Depressors of Scapula
(de•**press**•ors)   (**skap**•yoo•lah)

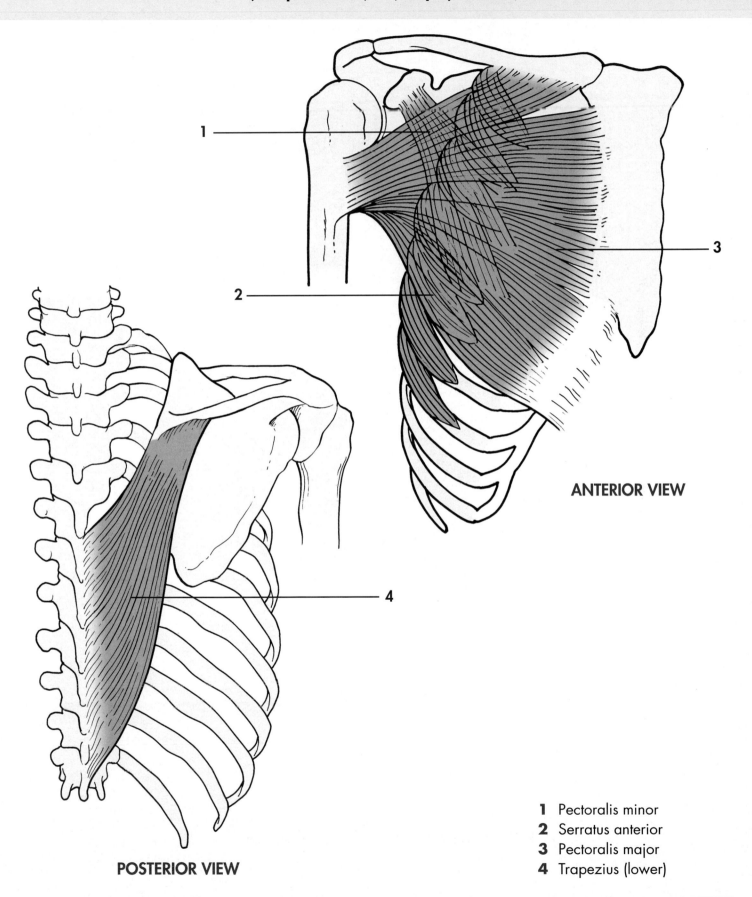

ANTERIOR VIEW

POSTERIOR VIEW

1 Pectoralis minor
2 Serratus anterior
3 Pectoralis major
4 Trapezius (lower)

Protractors of Scapula
(pro•**track**•tors) (**skap**•yoo•lah)

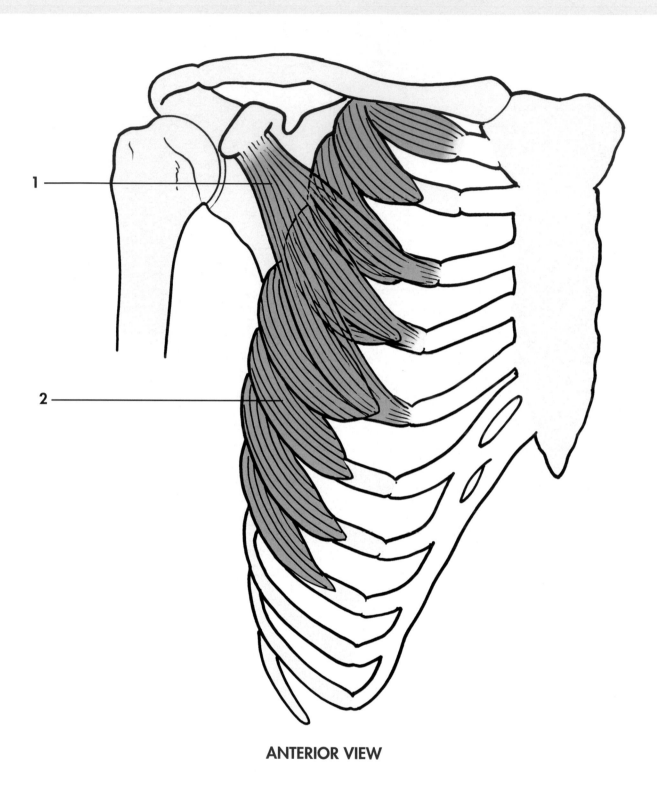

ANTERIOR VIEW

1 Pectoralis minor   **2** Serratus anterior

An Illustrated Atlas of the Skeletal Muscles

Retractors of Scapula
(re•**track**•tors) (**skap**•yoo•lah)

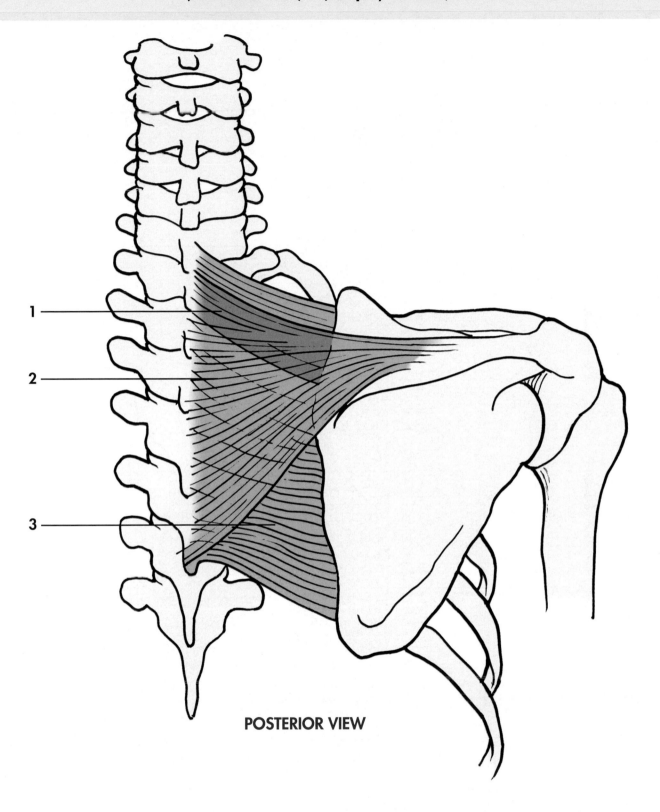

POSTERIOR VIEW

1 Rhomboideus minor   **2** Trapezius (middle)   **3** Rhomboideus major

Upward Rotators of Scapula
(ro•**tay**•tors) (**skap**•yoo•lah)

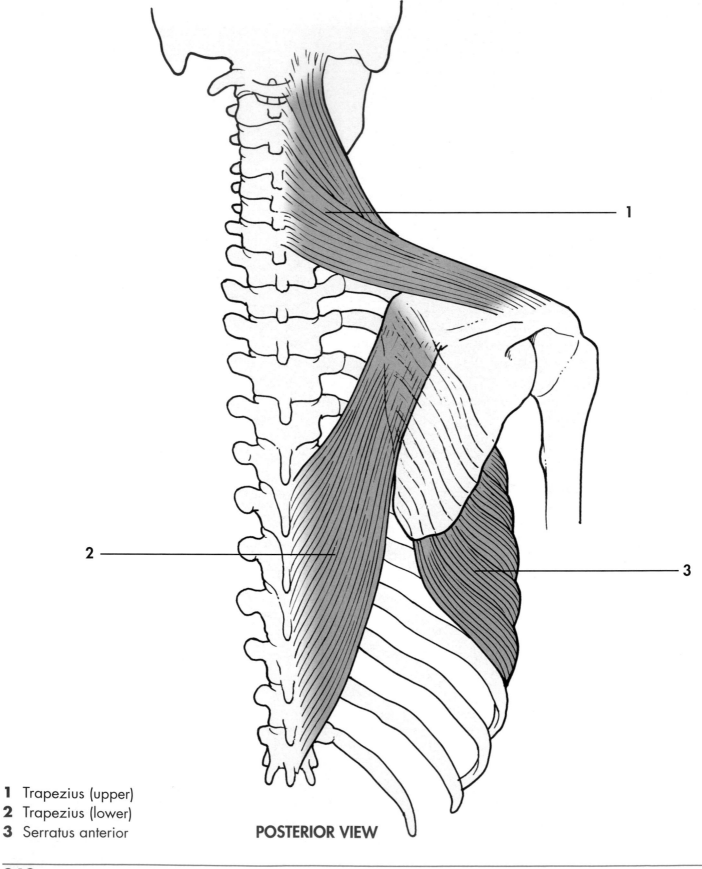

POSTERIOR VIEW

1 Trapezius (upper)
2 Trapezius (lower)
3 Serratus anterior

An Illustrated Atlas of the Skeletal Muscles

Downward Rotators of Scapula
(ro•**tay**•tors)   (**skap**•yoo•lah)

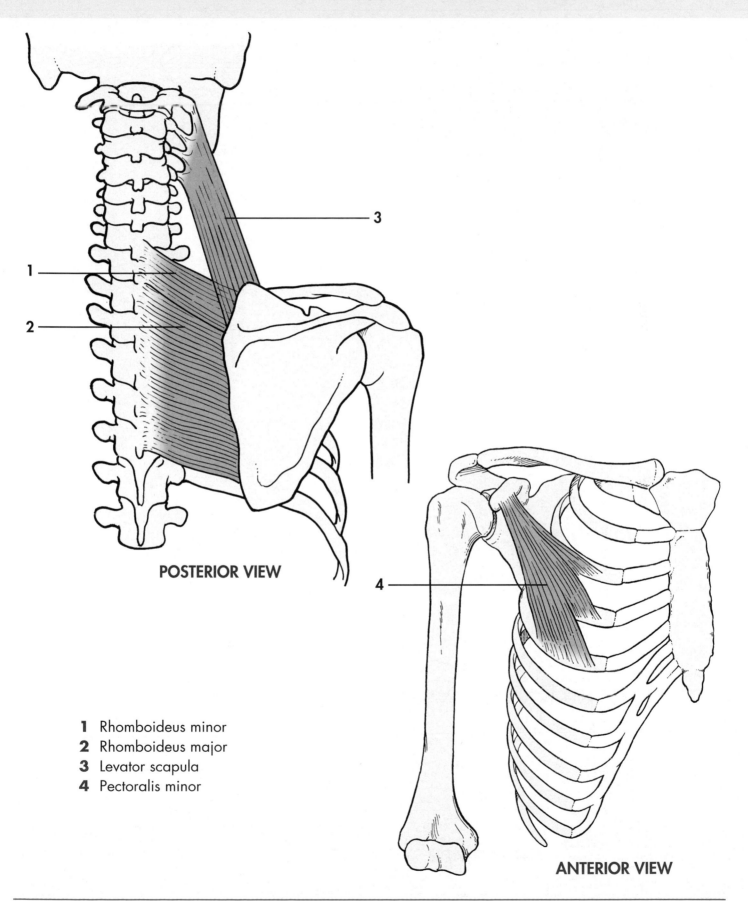

POSTERIOR VIEW

ANTERIOR VIEW

1 Rhomboideus minor
2 Rhomboideus major
3 Levator scapula
4 Pectoralis minor

Medial Rotators of Humerus

(ro•**tay**•tors) (**hyoo**•mir•us)

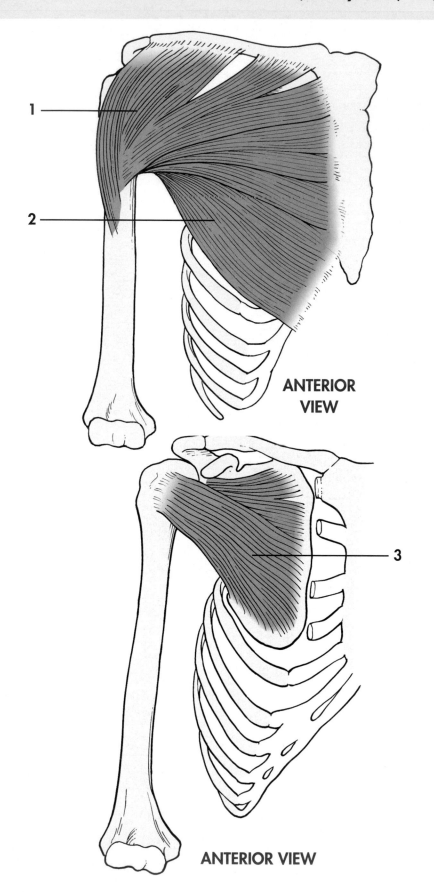

ANTERIOR VIEW

ANTERIOR VIEW

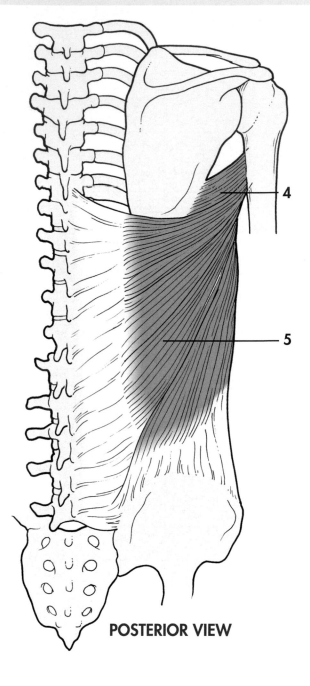

POSTERIOR VIEW

1 Deltoid (anterior)
2 Pectoralis major
3 Subscapularis
4 Teres major
5 Latissimus dorsi

An Illustrated Atlas of the Skeletal Muscles

Lateral Rotators of Humerus
(ro•**tay**•tors)   (**hyoo**•mir•us)

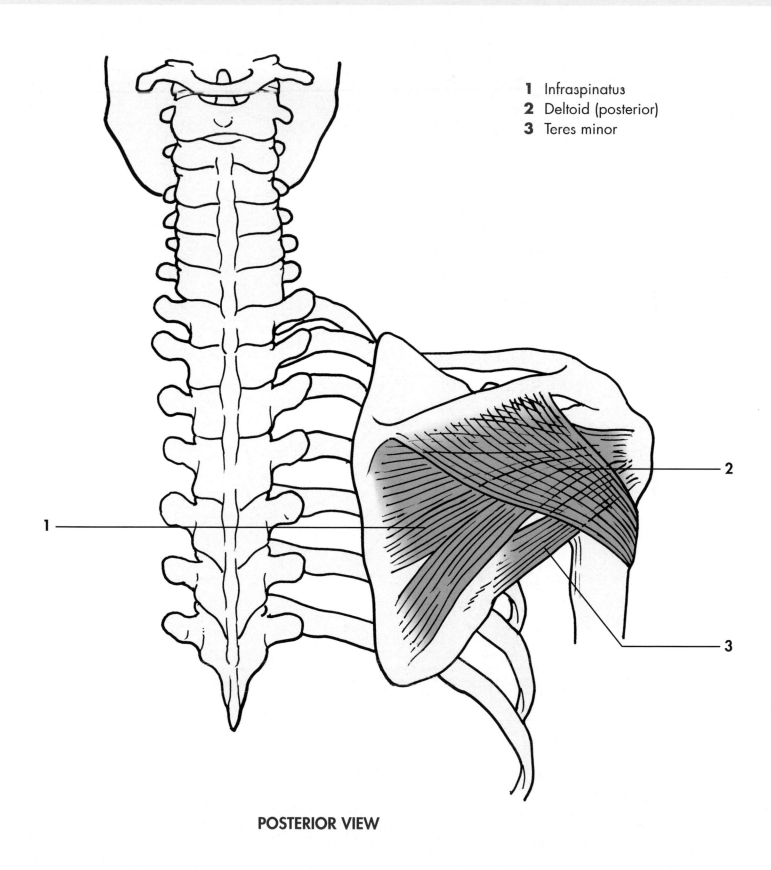

1 Infraspinatus
2 Deltoid (posterior)
3 Teres minor

POSTERIOR VIEW

Flexors of Humerus
(**flek**•sors)   (**hyoo**•mir•us)

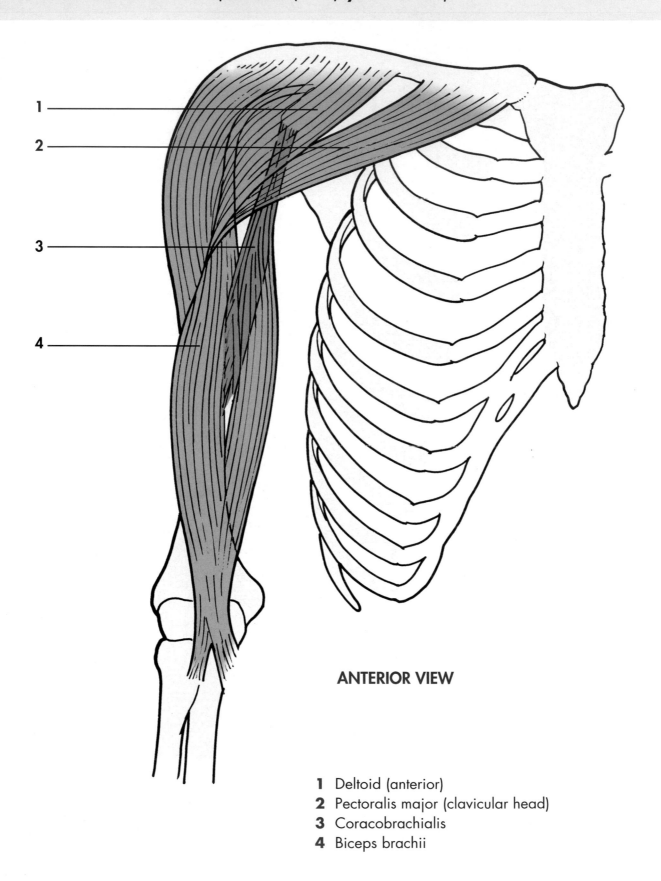

ANTERIOR VIEW

1 Deltoid (anterior)
2 Pectoralis major (clavicular head)
3 Coracobrachialis
4 Biceps brachii

Extensors of Humerus

(ex•**sten**•sors) (**hyoo**•mir•us)

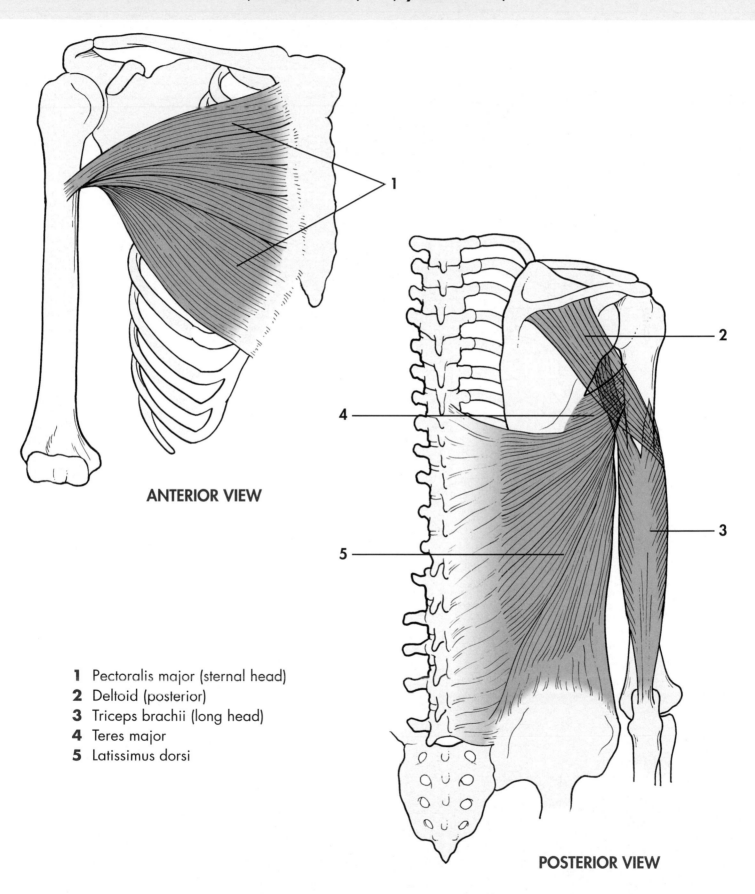

ANTERIOR VIEW

POSTERIOR VIEW

1 Pectoralis major (sternal head)
2 Deltoid (posterior)
3 Triceps brachii (long head)
4 Teres major
5 Latissimus dorsi

Abductors of Humerus
(ab•duck•tors)   (hyoo•mir•us)

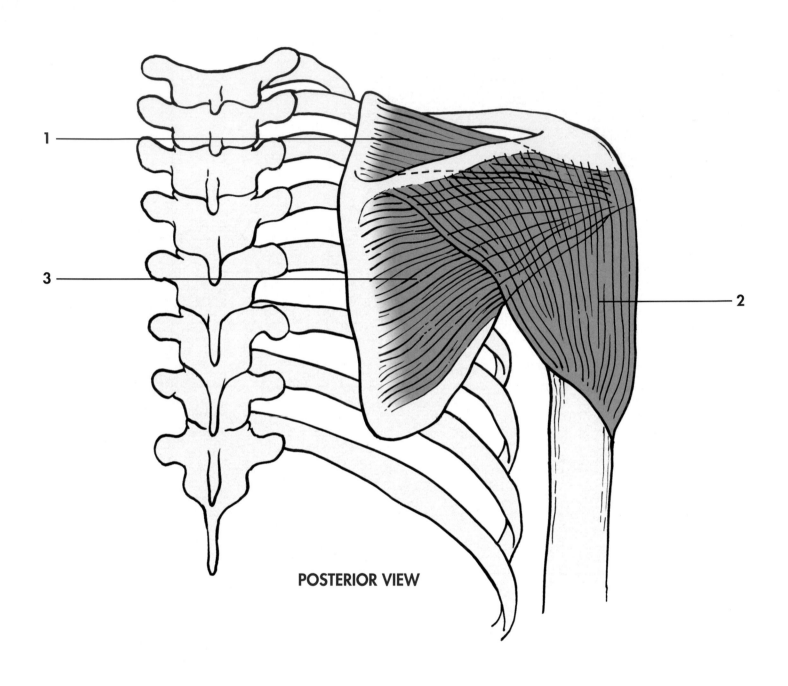

POSTERIOR VIEW

1 Supraspinatus
2 Deltoid (middle and posterior)
3 Infraspinatus

Adductors of Humerus
(ad•duck•tors) (hyoo•mir•us)

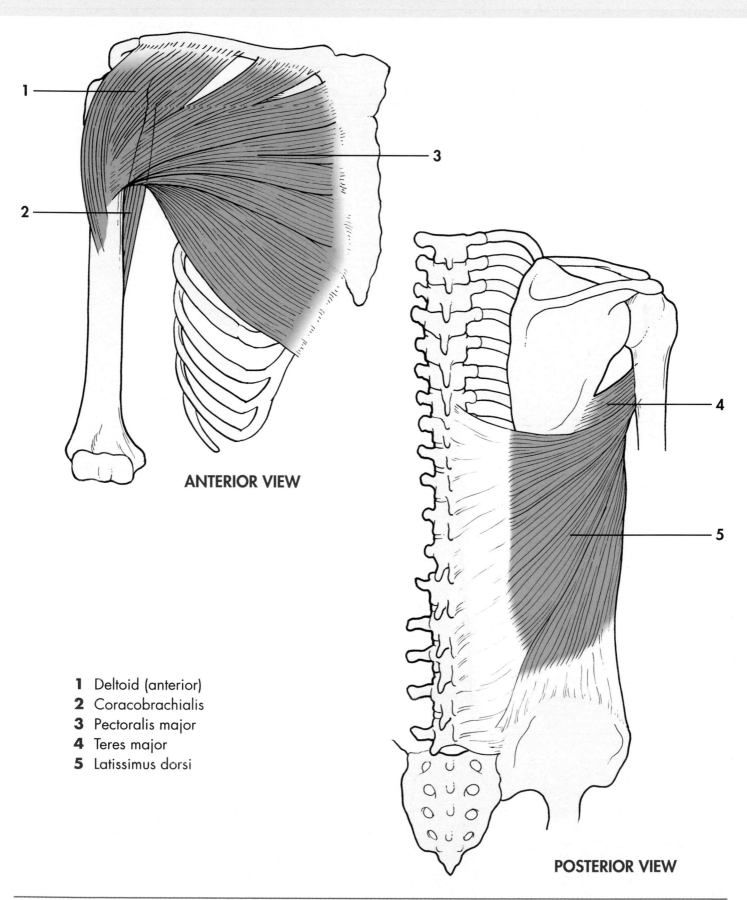

ANTERIOR VIEW

POSTERIOR VIEW

1 Deltoid (anterior)
2 Coracobrachialis
3 Pectoralis major
4 Teres major
5 Latissimus dorsi

Flexors of Elbow
(flek•sors)

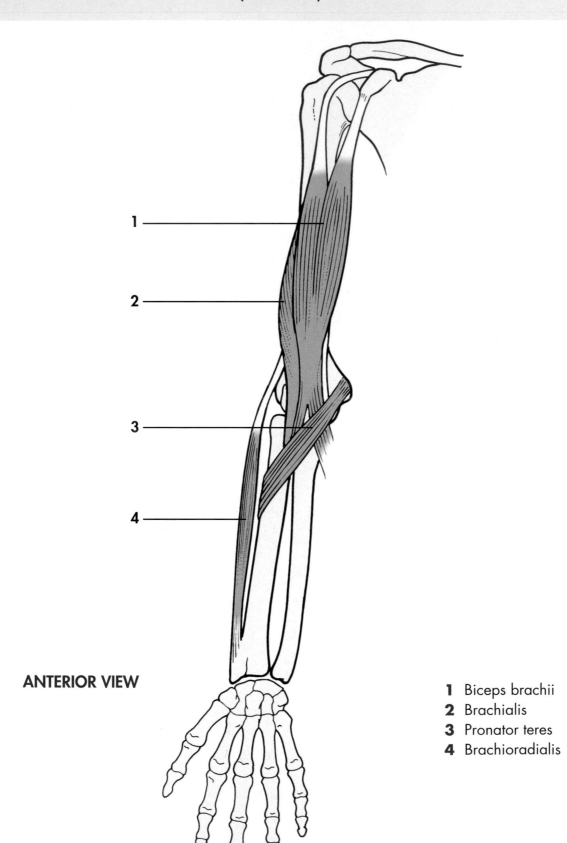

ANTERIOR VIEW

1 Biceps brachii
2 Brachialis
3 Pronator teres
4 Brachioradialis

An Illustrated Atlas of the Skeletal Muscles

Extensors of Elbow

(ex•**sten**•sors)

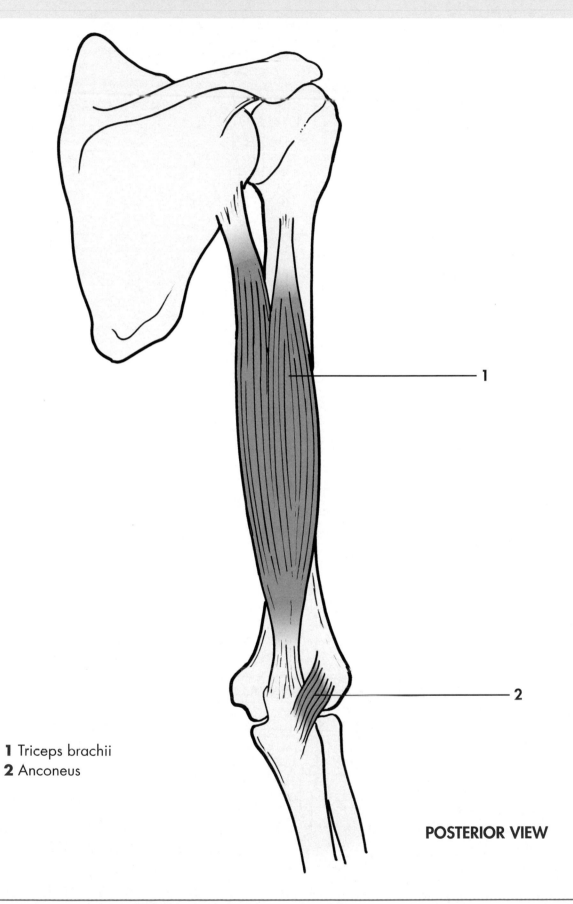

1 Triceps brachii
2 Anconeus

POSTERIOR VIEW

Supinators of Forearm
(soo•pin•**nay**•tors)

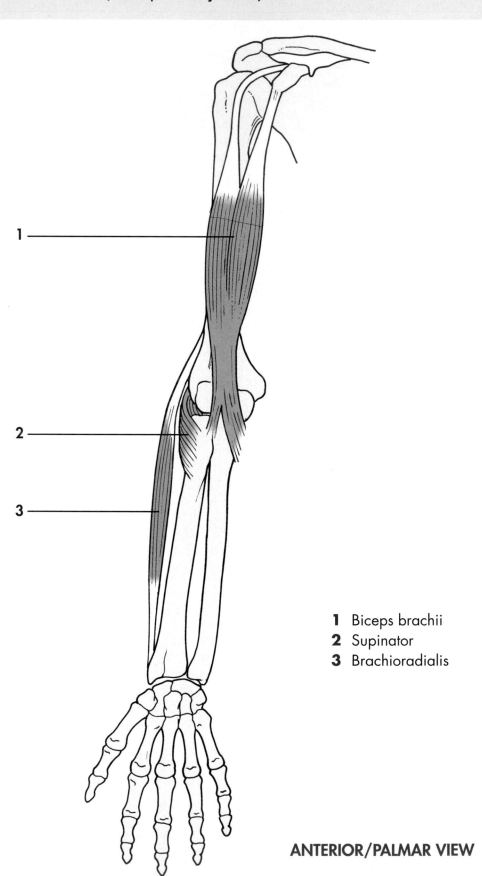

1 Biceps brachii
2 Supinator
3 Brachioradialis

ANTERIOR/PALMAR VIEW

An Illustrated Atlas of the Skeletal Muscles

Pronators of Forearm

(pro•**nay**•tors)

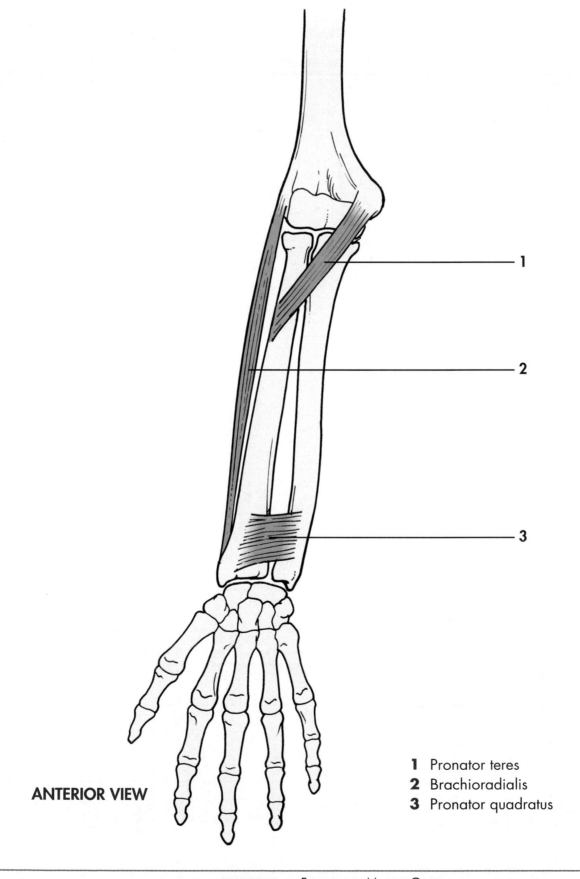

ANTERIOR VIEW

1 Pronator teres
2 Brachioradialis
3 Pronator quadratus

Flexors of Wrist
(flek•sors)

1 Flexor carpi radialis
2 Palmaris longus
3 Flexor carpi ulnaris
4 Flexor digitorum superficialis
5 Flexor digitorum profundus

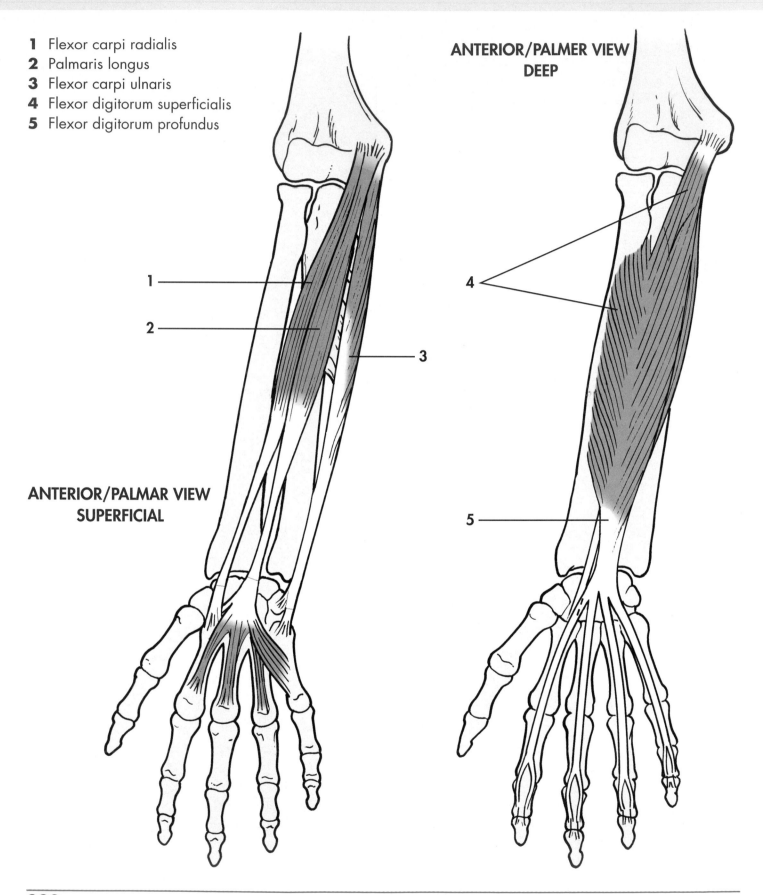

ANTERIOR/PALMER VIEW DEEP

ANTERIOR/PALMAR VIEW SUPERFICIAL

An Illustrated Atlas of the Skeletal Muscles

Extensors of Wrist

(ex•**sten**•sors)

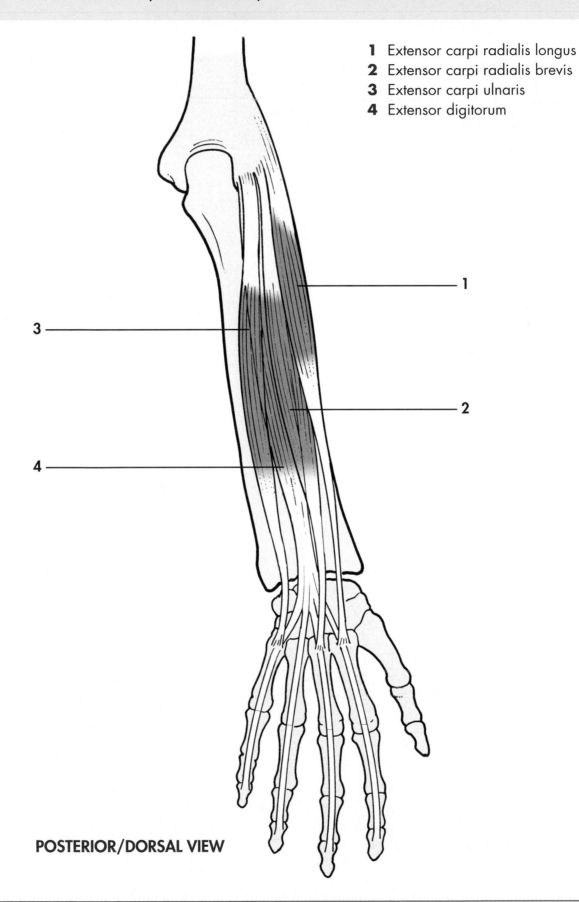

1 Extensor carpi radialis longus
2 Extensor carpi radialis brevis
3 Extensor carpi ulnaris
4 Extensor digitorum

POSTERIOR/DORSAL VIEW

Adductors of Wrist
(**ad**•duck•tors)

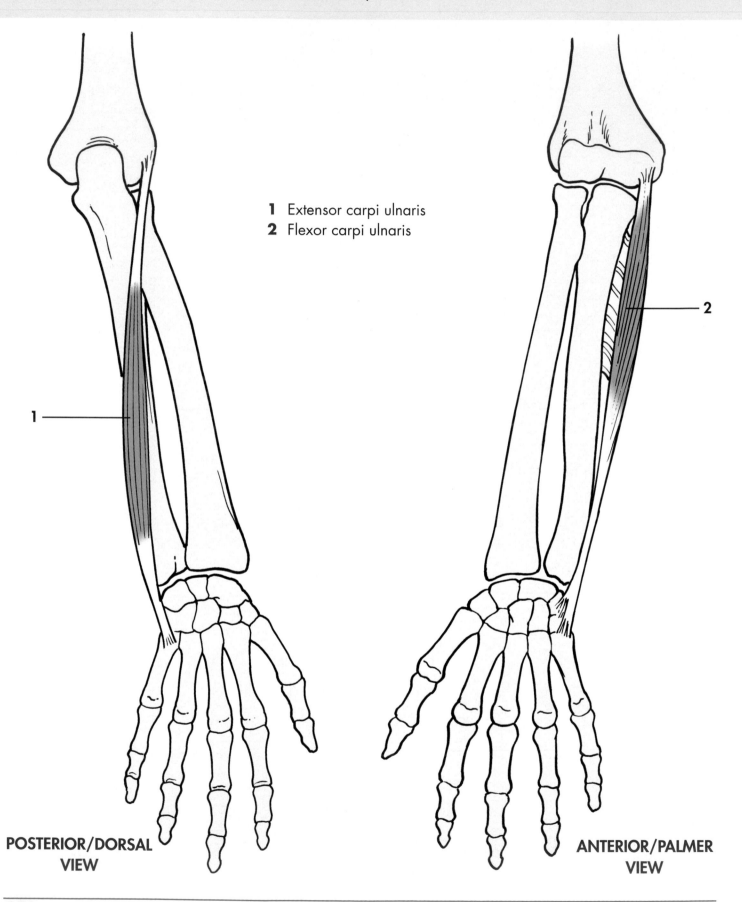

1 Extensor carpi ulnaris
2 Flexor carpi ulnaris

1

2

**POSTERIOR/DORSAL
VIEW**

**ANTERIOR/PALMER
VIEW**

An Illustrated Atlas of the Skeletal Muscles

Abductors of Wrist

(**ab**•duck•tors)

1 Extensor carpi radialis longus
2 Extensor carpi radialis brevis
3 Abductor pollicis longus
4 Flexor carpi radialis
5 Flexor pollicis longus

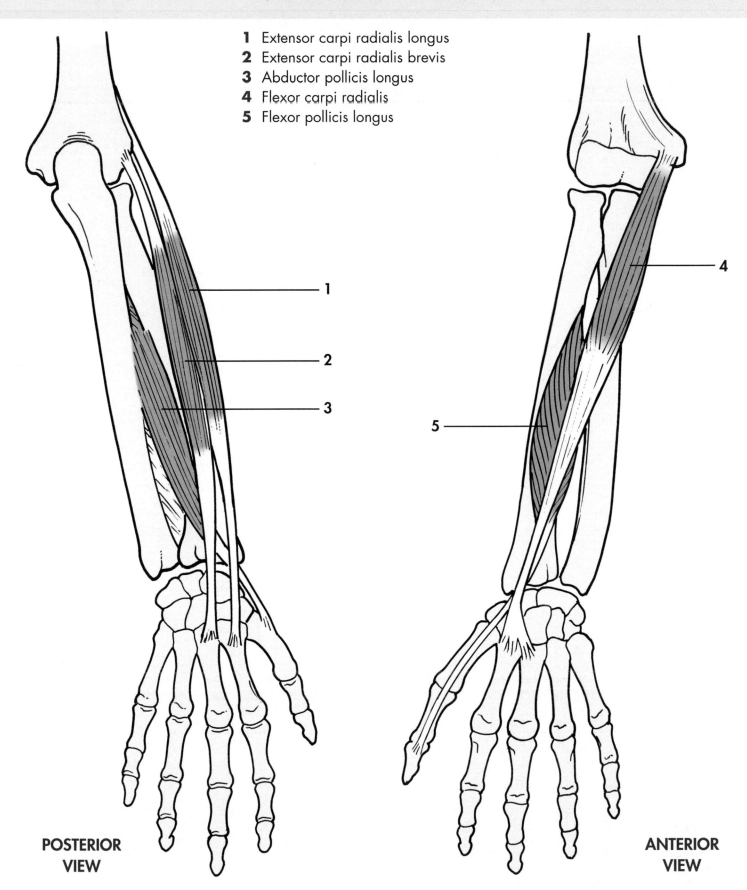

POSTERIOR
VIEW

ANTERIOR
VIEW

Abductors of Thumb and Digits
(**ab**•duck•tors)

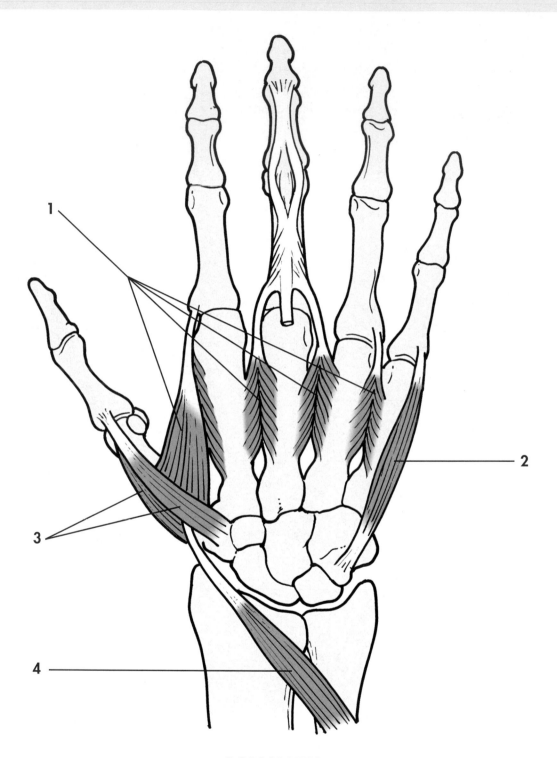

DORSAL VIEW

Fingers
1 Dorsal interossei
2 Abductor digiti minimi

Thumb
3 Abductor pollicis brevis
4 Abductor pollicis longus

An Illustrated Atlas of the Skeletal Muscles

Adductors of Thumb and Digits
(**ad**•duck•tors)

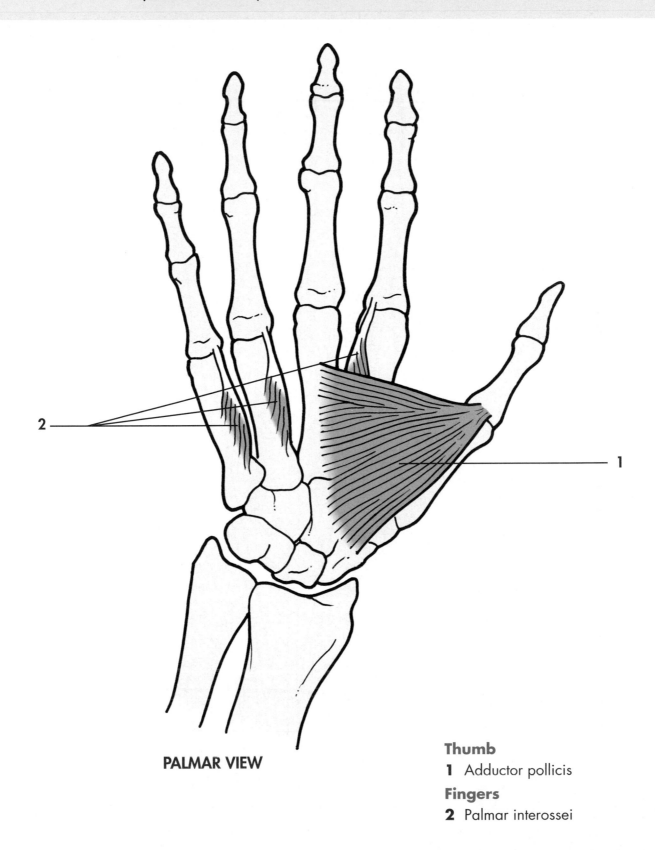

2

1

PALMAR VIEW

Thumb
1 Adductor pollicis

Fingers
2 Palmar interossei

Extension-Opposition of Thumb and Digits
(ex•**sten**•shun)(**op**•poh•zih•shun)

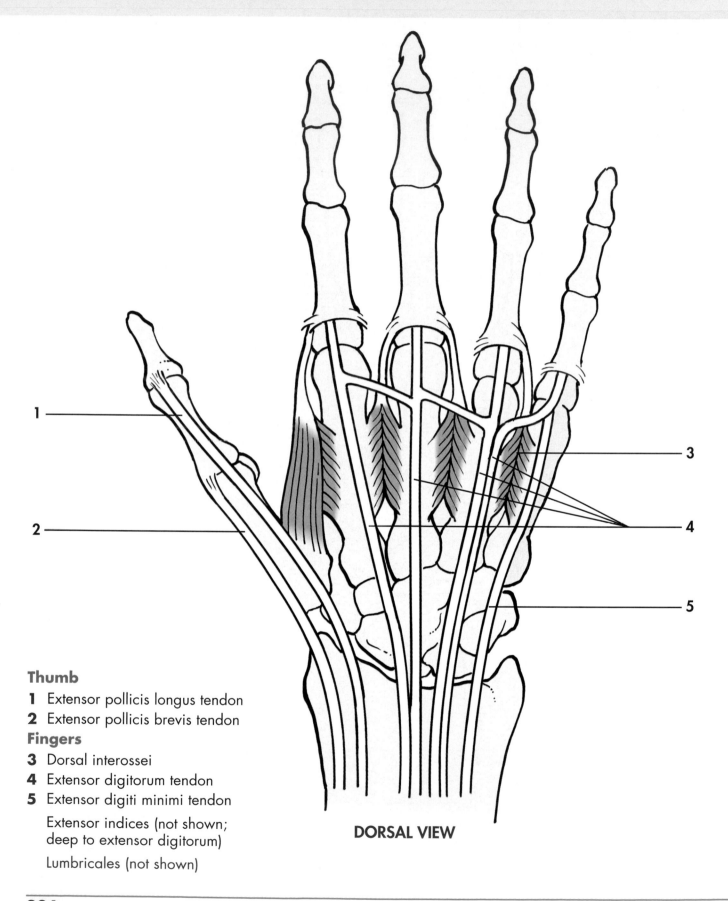

DORSAL VIEW

Thumb
1 Extensor pollicis longus tendon
2 Extensor pollicis brevis tendon
Fingers
3 Dorsal interossei
4 Extensor digitorum tendon
5 Extensor digiti minimi tendon

Extensor indices (not shown; deep to extensor digitorum)

Lumbricales (not shown)

An Illustrated Atlas of the Skeletal Muscles

Flexion-Opposition of Thumb and Digits
(**flek**•shun) (**op**•poh•zih•shun)

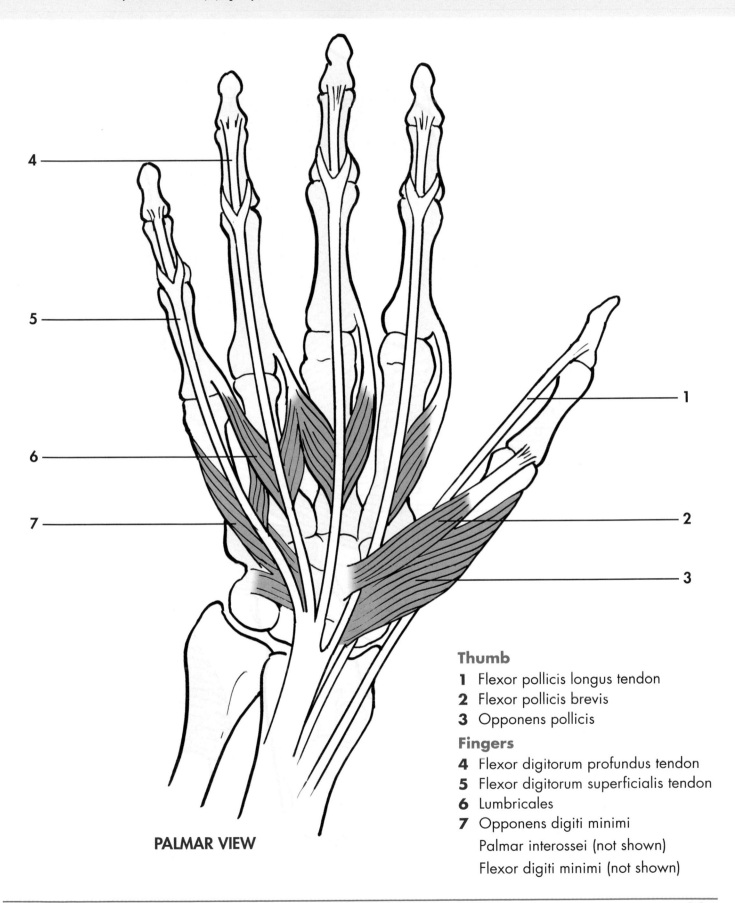

PALMAR VIEW

Thumb
1 Flexor pollicis longus tendon
2 Flexor pollicis brevis
3 Opponens pollicis

Fingers
4 Flexor digitorum profundus tendon
5 Flexor digitorum superficialis tendon
6 Lumbricales
7 Opponens digiti minimi
  Palmar interossei (not shown)
  Flexor digiti minimi (not shown)

Extensors of Hip
(ex•**sten**•sors)

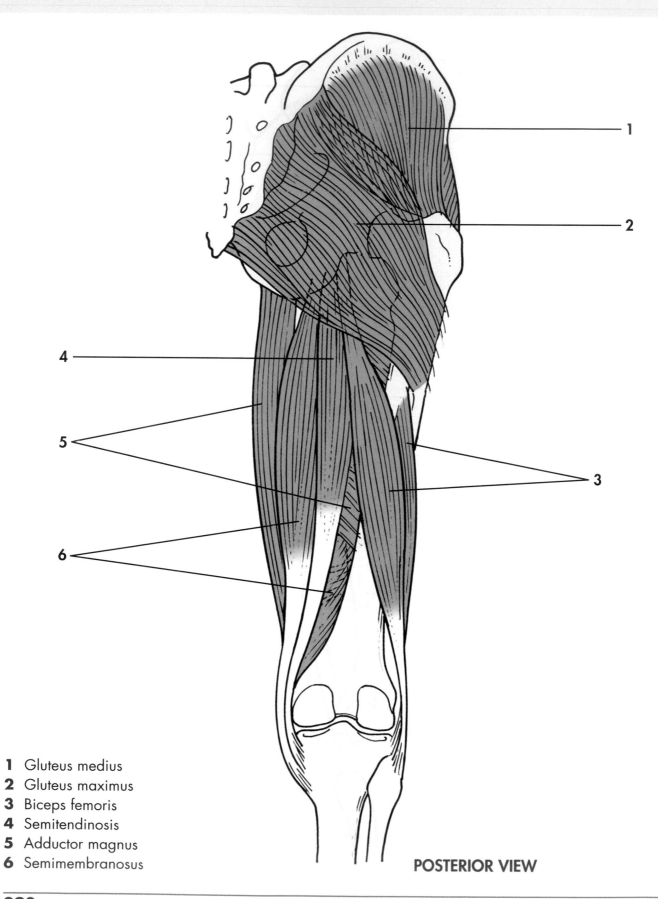

1 Gluteus medius
2 Gluteus maximus
3 Biceps femoris
4 Semitendinosis
5 Adductor magnus
6 Semimembranosus

POSTERIOR VIEW

An Illustrated Atlas of the Skeletal Muscles

Flexors of Hip
(flek•sors)

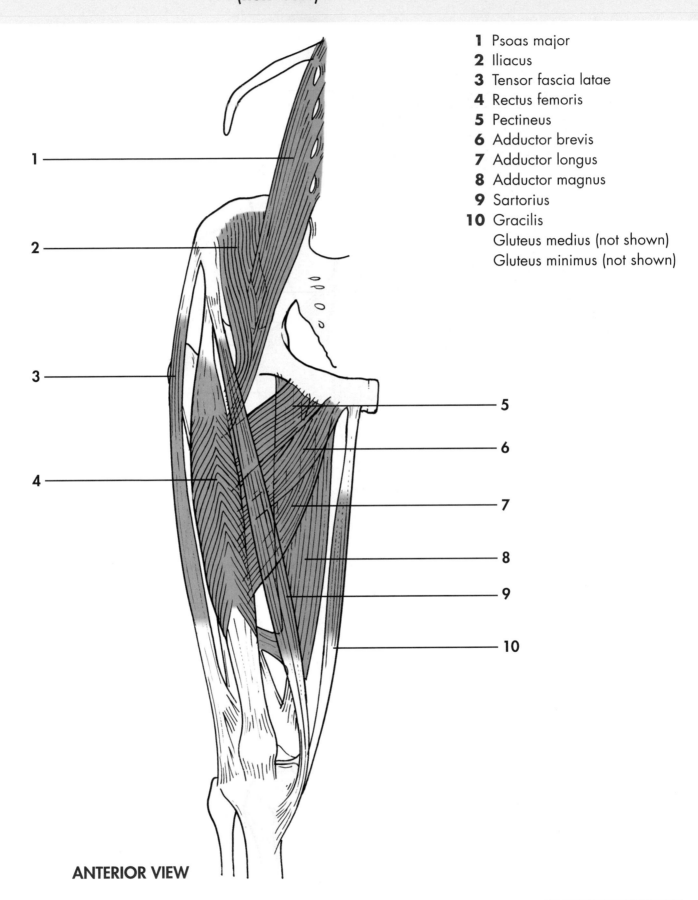

1 Psoas major
2 Iliacus
3 Tensor fascia latae
4 Rectus femoris
5 Pectineus
6 Adductor brevis
7 Adductor longus
8 Adductor magnus
9 Sartorius
10 Gracilis
   Gluteus medius (not shown)
   Gluteus minimus (not shown)

ANTERIOR VIEW

Lateral Rotators of Knee
(ro•**tay**•tors)

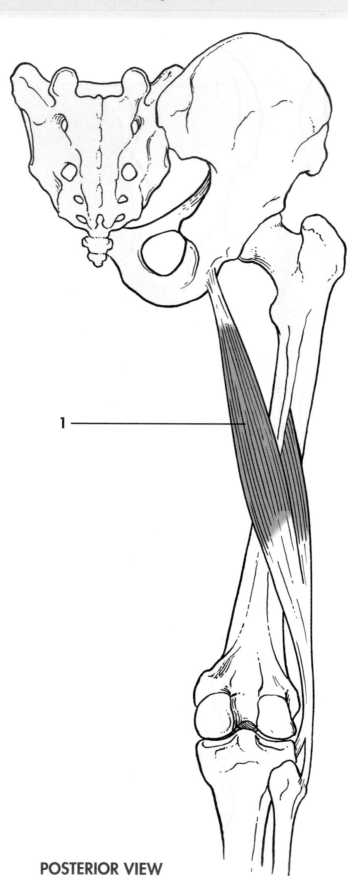

1 ———

1 Biceps femoris

POSTERIOR VIEW

An Illustrated Atlas of the Skeletal Muscles

Medial Rotators of Knee
(ro•**tay**•tors)

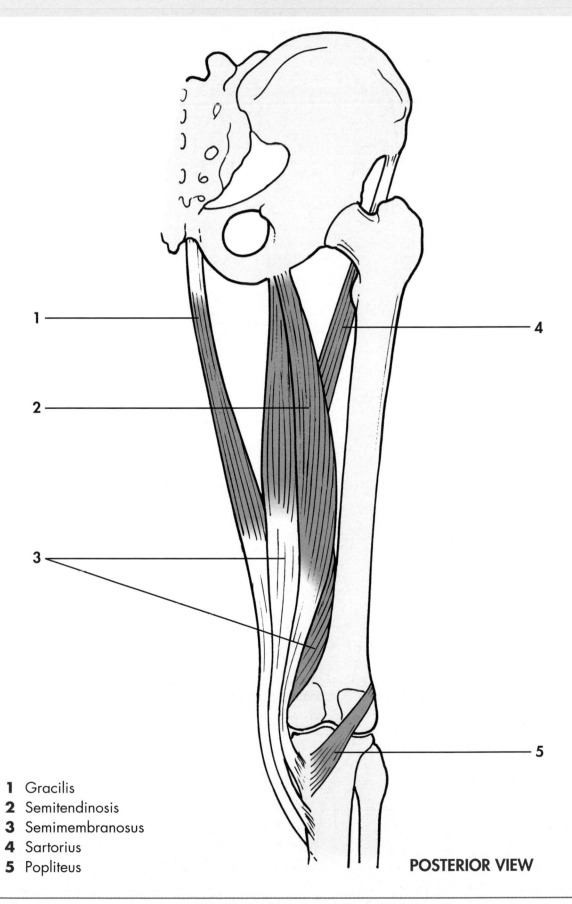

1
2
3
4
5

POSTERIOR VIEW

1 Gracilis
2 Semitendinosis
3 Semimembranosus
4 Sartorius
5 Popliteus

Extensors of Knee

(ex•**sten**•sors)

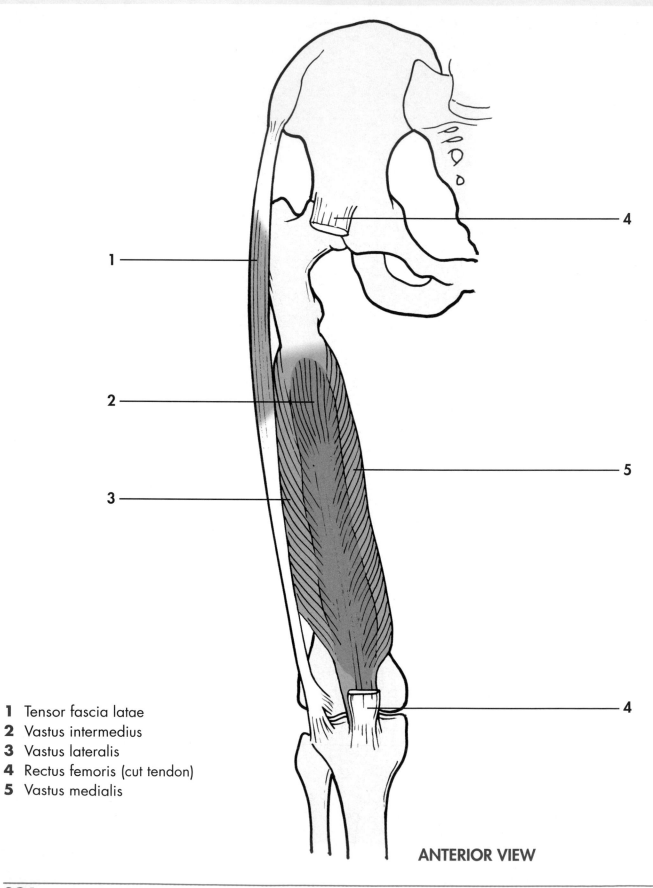

1 Tensor fascia latae
2 Vastus intermedius
3 Vastus lateralis
4 Rectus femoris (cut tendon)
5 Vastus medialis

ANTERIOR VIEW

An Illustrated Atlas of the Skeletal Muscles

Flexors of Knee
(**flek**•sors)

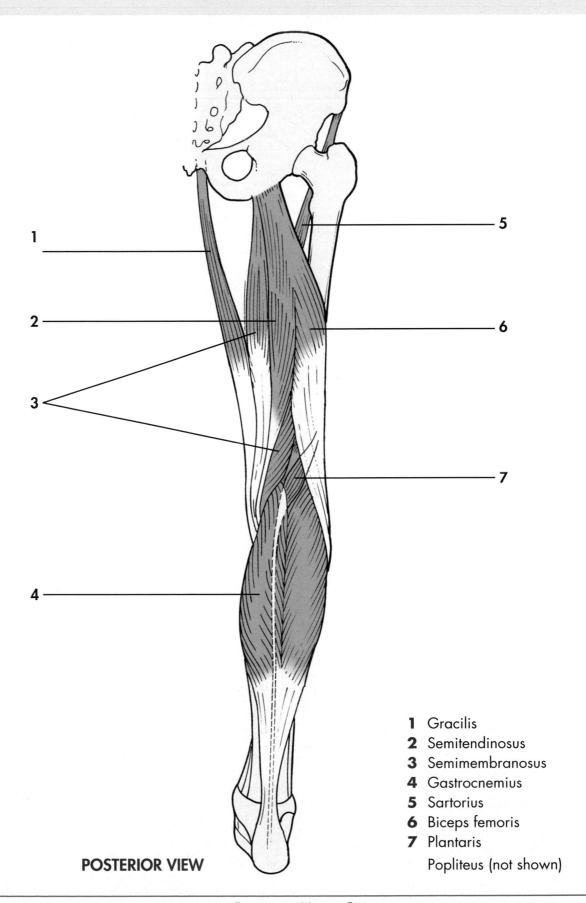

1 Gracilis
2 Semitendinosus
3 Semimembranosus
4 Gastrocnemius
5 Sartorius
6 Biceps femoris
7 Plantaris
Popliteus (not shown)

POSTERIOR VIEW

Dorsiflexors of Ankle
(door•se•**flek**•sors)

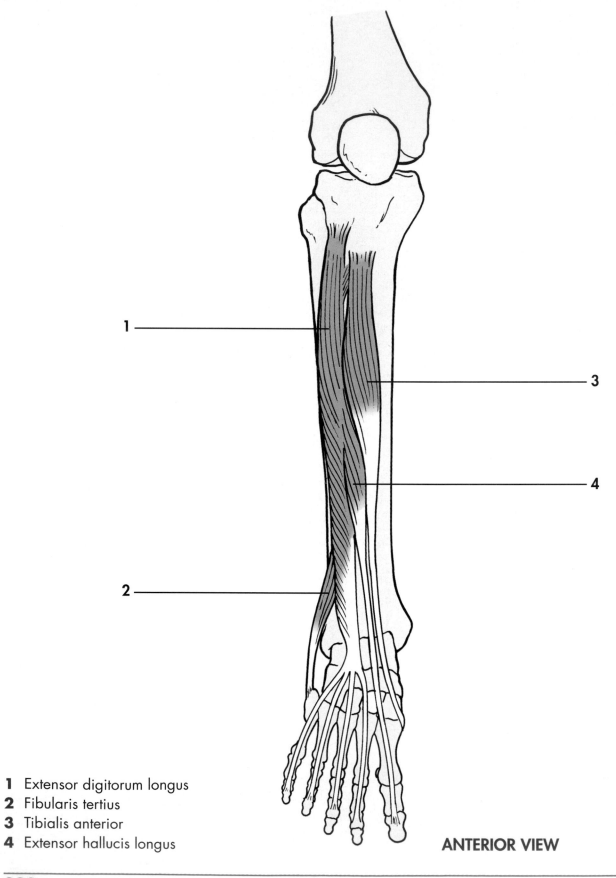

1   Extensor digitorum longus
2   Fibularis tertius
3   Tibialis anterior
4   Extensor hallucis longus

ANTERIOR VIEW

An Illustrated Atlas of the Skeletal Muscles

Plantar Flexors of Ankle
(plan•tahr) (flek•sors)

1 Plantaris
2 Gastrocnemius
3 Flexor hallucis longus
4 Fibularis longus
5 Fibularis brevis
6 Flexor digitorum longus
7 Soleus
   Tibialis posterior (not shown)

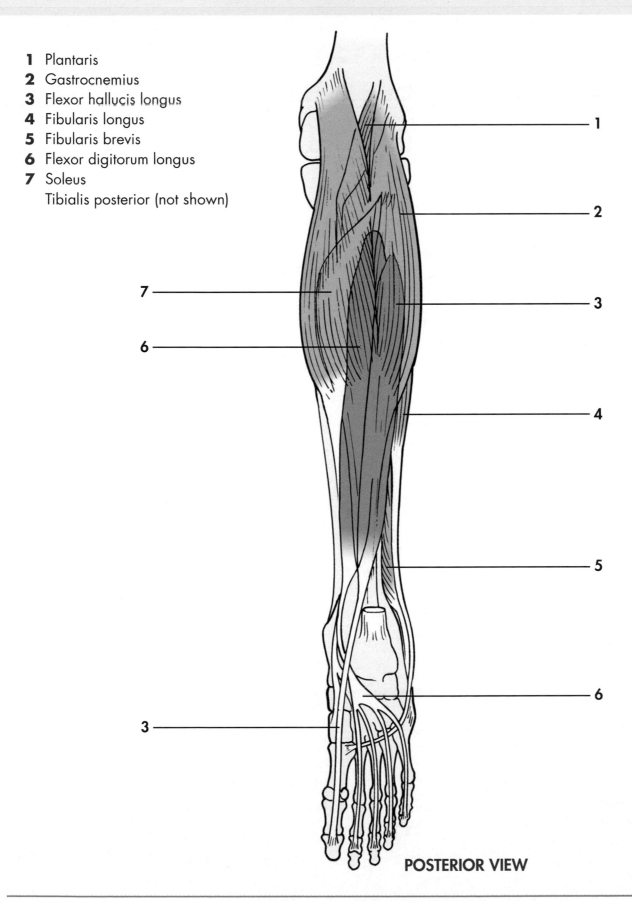

POSTERIOR VIEW

Invertors of Foot
(in•**ver**•tors)

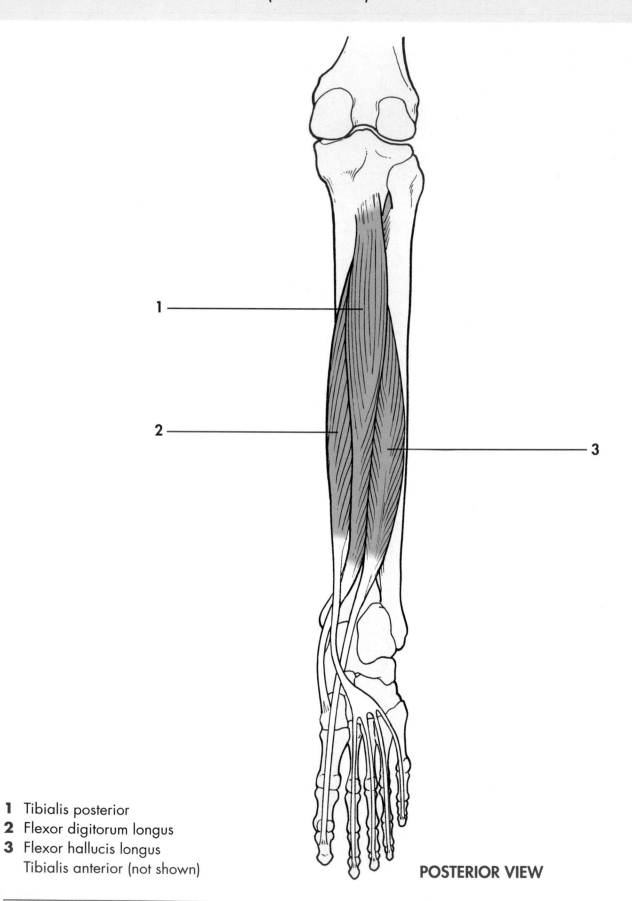

1 ————

2 ————

3 ————

1 Tibialis posterior
2 Flexor digitorum longus
3 Flexor hallucis longus
Tibialis anterior (not shown)

POSTERIOR VIEW

An Illustrated Atlas of the Skeletal Muscles

Evertors of Foot
(ee•**ver**•tors)

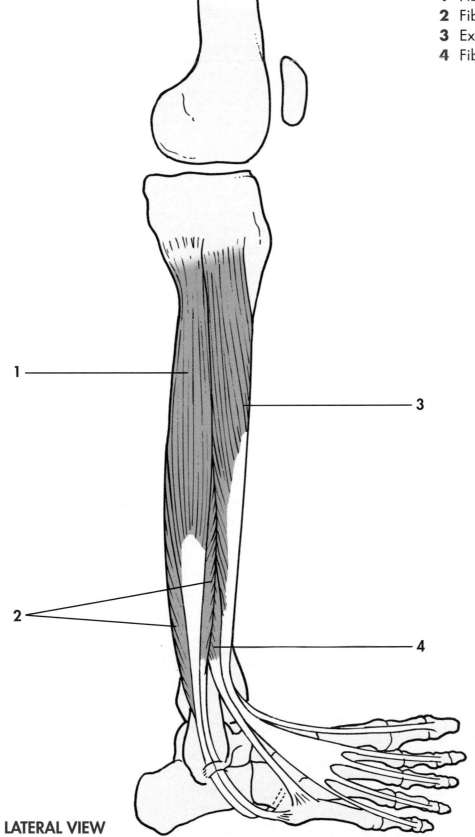

1 Fibularis longus
2 Fibularis brevis
3 Extensor digitorum longus
4 Fibularis tertius

LATERAL VIEW

Muscle Innervation Pathways

12

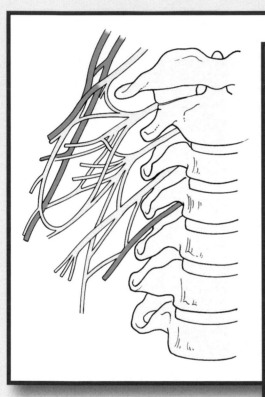

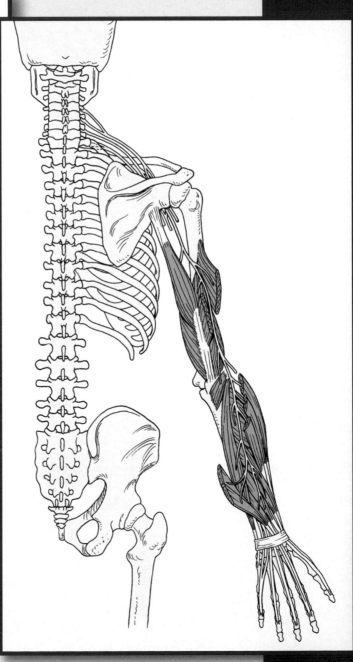

Cervical Plexus—Roots and Primary Branches

(**ser**•vi•kal) (**plex**•us)

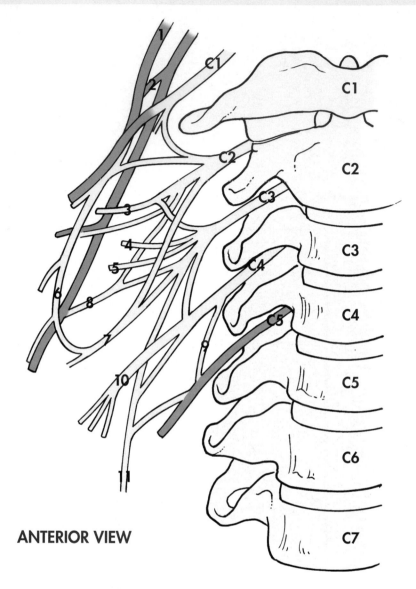

ANTERIOR VIEW

1 Hypoglossal nerve (XII)
2 Cranial branch of spinoaccessory nerve (XI)
3 Lesser occipital nerve
4 Great auricular nerve
5 Transverse cervical nerve
6 Superior root of ansa cervicalis

7 Inferior root of ansa cervicalis
8 Spinal branch to spinoaccessory nerve (XI)
9 Branch to brachial plexus
10 Supraclavicular nerves
11 Phrenic nerve

The **cervical plexus** is formed by the anterior branches of the first four cervical nerves (C1–C4) and a portion of C5. The cervical plexus branches innervate the skin and muscles of the neck and areas of the head and shoulders. Some cervical plexus fibers join with the **accessory and hypoglossal cranial nerves** to innervate specific neck and pharyngeal muscles. Nerve fibers from the third, fourth and fifth cervical nerves combine to form the **phrenic nerve**, which innervates the diaphragm causing it to contract and flatten, drawing air into the lungs.

An Illustrated Atlas of the Skeletal Muscles

Cervical Plexus—Peripheral Innervations
(**ser**•vi•kal) (**plex**•us)

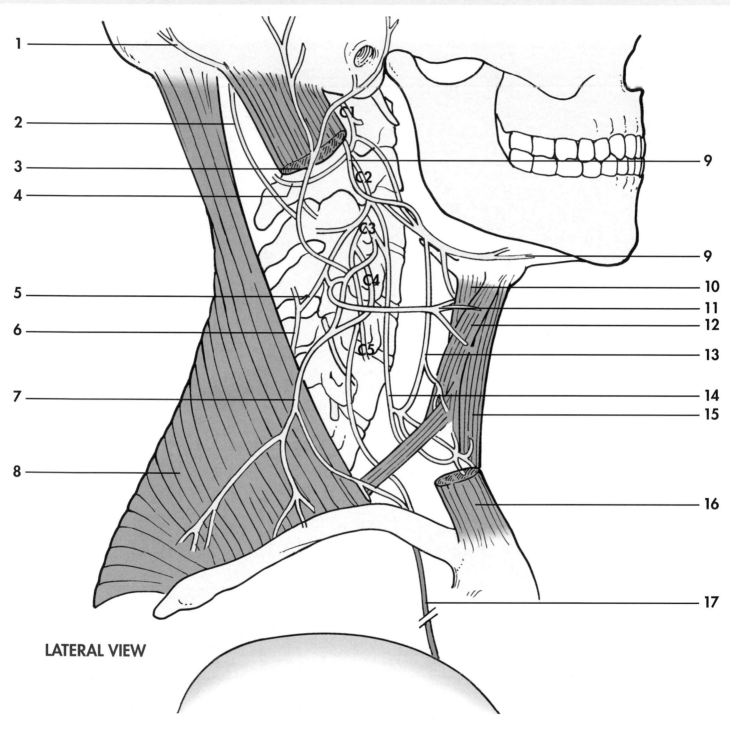

LATERAL VIEW

1 Greater occipital nerve
2 Lesser occipital nerve
3 Greater auricular nerve
4 Suboccipital nerve (C1)
5 Spinal branch of spinoaccessory nerve (XI)

6 Spinoaccessory nerve
(CN XI–cut)
7 Supraclavicular nerves
8 Trapezius muscle
9 Hypoglossal nerve (CN XII)
10 Thyrohyoid muscle
11 Transverse cervical nerve

12 Omohyoid muscle
13 Superior root of ansa cervicalis
14 Inferior root of ansa cervicalis
15 Sternohyoid muscle
16 Sternocleidomastoid muscle
17 Phrenic nerve

Brachial Plexus and Axillary (Circumflex) Nerve

(bray•key•al) (**plex**•us) (**ax**•i•ler•ee) (**sir**•kum•flex)

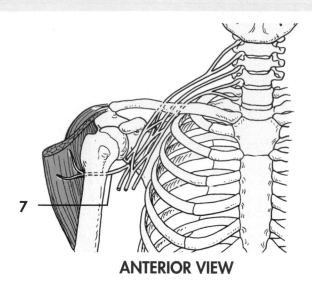

ANTERIOR VIEW

The **axillary nerve** innervates the **deltoid** and **teres minor** muscles. Those muscles are major movers of the shoulder. The skin in this region is also innervated by this nerve.

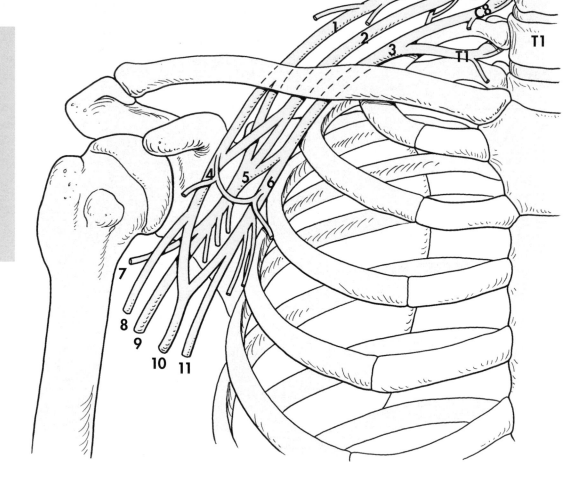

1 Superior trunk	**5** Posterior cord	**9** Radial nerve
2 Middle trunk	**6** Medial cord	**10** Median nerve
3 Inferior trunk	**7** Axillary nerve	**11** Ulnar nerve
4 Lateral cord	**8** Musculocutaneous nerve	

Musculocutaneous Nerve
(mus•kew•lo•kew•tay•ne•us)

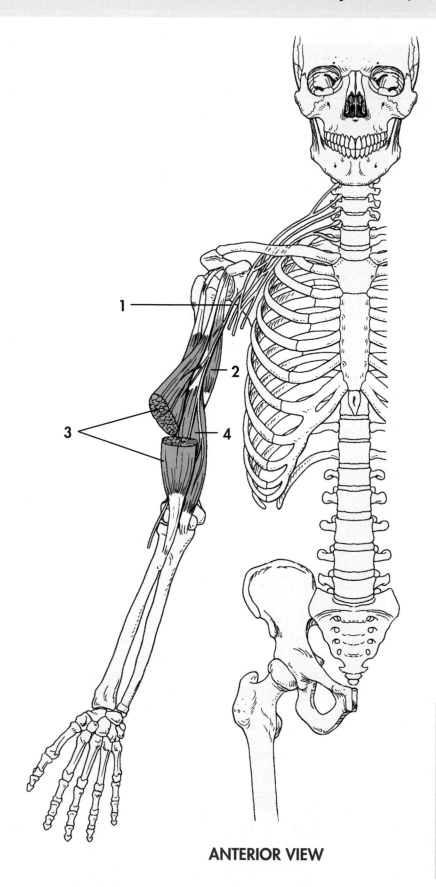

1 Musculocutaneous nerve
2 Coracobrachialis
3 Biceps brachii
4 Brachialis

The **musculocutaneous nerve** innervates the **coracobrachialis**, **biceps brachii**, and **brachialis** muscles. These muscles are major flexors of the forearm. It also innervates the skin of the anterior surface of the upper arm and forearm.

ANTERIOR VIEW

Radial Nerve
(ray•de•al)

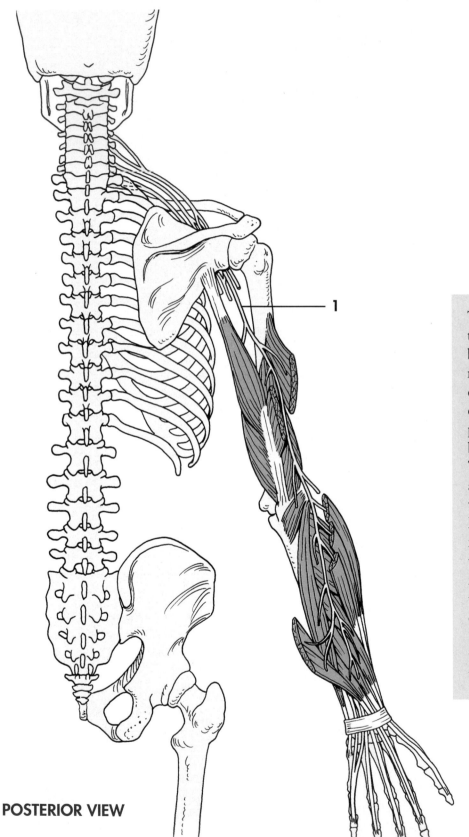

POSTERIOR VIEW

1 Radial nerve

The **radial** nerve innervates the **triceps brachii, anconeus, brachioradialis, extensor carpi radialis, supinator, extensor digitorum, extensor digiti minimi, extensor carpi ulnaris, abductor pollicis longus, extensor pollicis brevis,** and **extensor indicis.** These muscles extend the forearm and the hand. It also innervates the skin of the dorsal forearm and hand. Crutches that are too long may injure the posterior cord of the brachial plexus, affecting the radial nerve and paralyzing the triceps brachii, anconeus, and extensor muscles of the lower arm, producing wrist drop and the inability to extend the elbow, wrist, or fingers.

Median Nerve
(mee•de•an)

1 Median nerve

The **median nerve** innervates the **pronator teres, flexor carpi radialis, palmaris longus, flexor digitorum superficialis, flexor pollicis brevis, flexor pollicis longus, pronator quadratus, abductor pollicis brevis,** and the **first** and **second lumbricals.** These muscles are flexors of the forearm and the hand. The skin of the palmar surface of the hand, thumb, index finger, and middle fingers are also innervated.

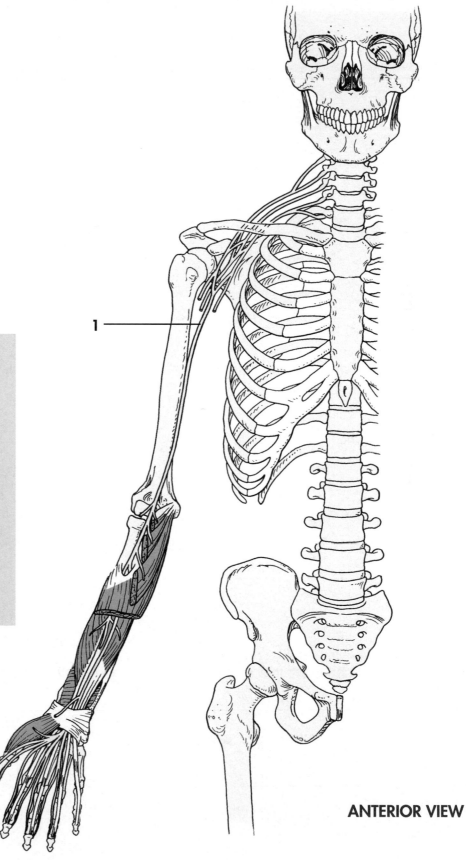

ANTERIOR VIEW

Ulnar Nerve
(ul•nar)

1 Ulnar nerve

The **ulnar nerve** innervates the **flexor carpi ulnaris, flexor digitorum profundus, adductor pollicis, flexor pollicis brevis, palmar interossei, abductor digiti minimi, opponens digiti minimi, dorsal interossei,** and the **third** and **fourth lumbricals**. These muscles flex the hand. It also innervates the skin of the anterior hand and the ring and little fingers.

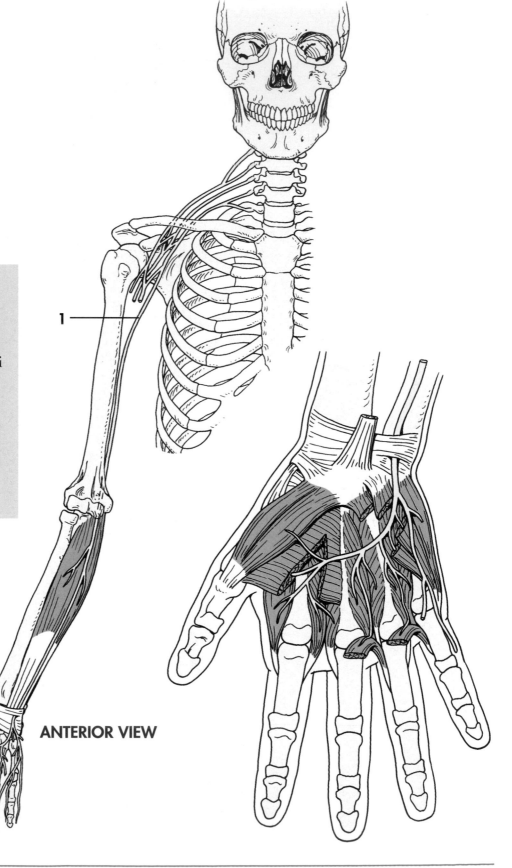

ANTERIOR VIEW

Lumbosacral Plexus

(lum•bo•**sa**•kral) (**plex**•us)

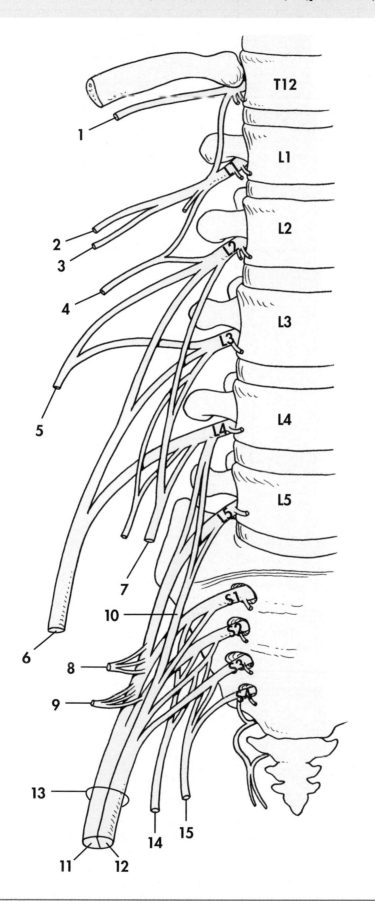

ANTERIOR VIEW

T12

L1

L2

L3

L4

L5

S1

S2

S3

S4

1 12th thoracic spinal nerve
2 Iliohypogastric nerve
3 Ilioinguinal nerve
4 Genitofemoral nerve
5 Lateral femoral cutaneous nerve
6 Femoral nerve
7 Obturator nerve
8 Superior gluteal nerve
9 Inferior gluteal nerve
10 Lumbosacral trunk
11 Common peroneal nerve
12 Tibial nerve
13 Sciatic nerve
14 Posterior femoral cutaneous nerve
15 Pudendal nerve

Obturator Nerve

(ob•too•**ray**•ter)

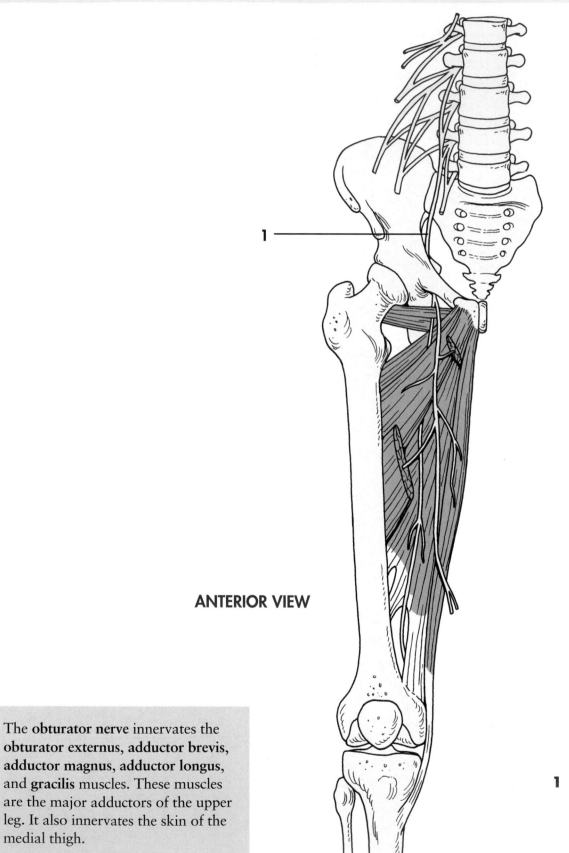

ANTERIOR VIEW

The **obturator nerve** innervates the **obturator externus, adductor brevis, adductor magnus, adductor longus,** and **gracilis** muscles. These muscles are the major adductors of the upper leg. It also innervates the skin of the medial thigh.

1 Obturator nerve

Femoral Nerve
(fem•er•al)

The **femoral nerve** innervates the **iliacus, sartorius, rectus femoris, pectineus, vastus lateralis,** and **vastus medialis** muscles. These muscles are hip flexors and extensors of the lower leg. It also innervates the skin of the anterior, lateral, and posterior portions of the thigh. The femoral nerve facilitates the **patellar reflex** or **"knee jerk response"** routinely tested during a physical examination. Tapping the **patellar ligament** with a "reflex" or percussion rubber hammer stimulates muscle spindle receptors in the quadriceps femoris muscle. Sensory (afferent) impulses are conveyed along the femoral nerve to the spinal cord (L2, L3, and L4 segments). Motor (efferent) impulses are transmitted via motor fibers in the femoral nerve to the quadriceps femoris, producing the "jerk-like" contraction of the muscle and extension of the leg at the knee joint.

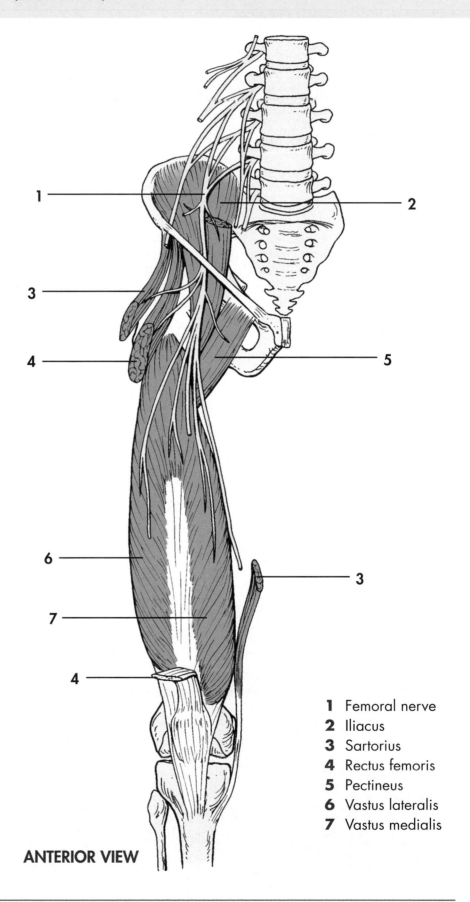

ANTERIOR VIEW

1 Femoral nerve
2 Iliacus
3 Sartorius
4 Rectus femoris
5 Pectineus
6 Vastus lateralis
7 Vastus medialis

Sciatic Nerve
(sy•at•tic)

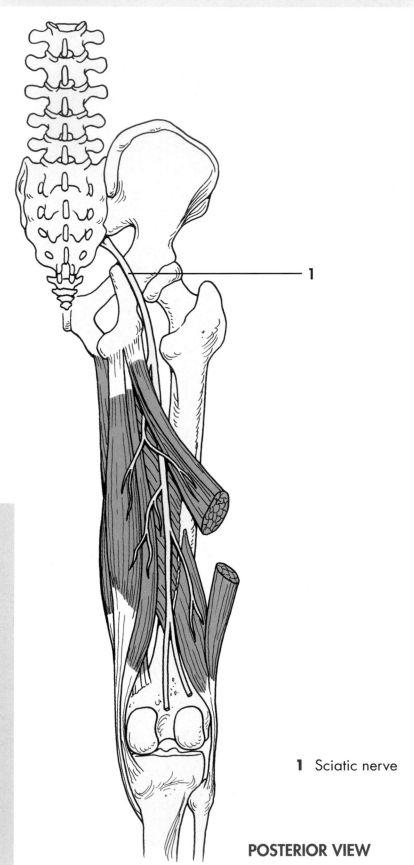

1

The **sciatic nerve** is composed of two divisions: the **tibial division** and the **fibular division**. The **tibial division** innervates the **long head of the biceps femoris**, the **semitendinosus**, the **semi-membranosus**, and the posterior portion of the **adductor magnus**. The **fibular division** innervates the **short head of the biceps femoris**. The **sciatic nerve** divides into the **tibial** and **fibular nerves** at the popliteal fossa. It also innervates the skin of the leg. "**Driver's thigh**" is a sciatic neuralgia caused by pressure on the gluteal branch resulting from the use of the accelerator in driving a car or truck long distances. Cruise control was designed to prevent this condition.

1 Sciatic nerve

POSTERIOR VIEW

Common Fibular (Peroneal) Nerve

(**fib**•yoo•lar) (perr•o•**nee**•al)

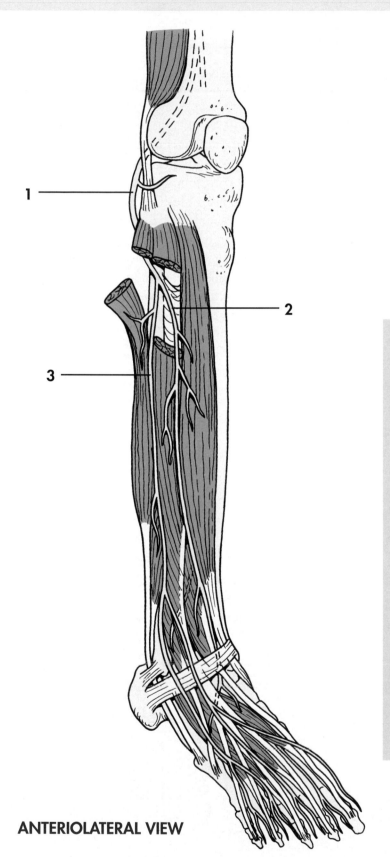

ANTERIOLATERAL VIEW

1 Common fibular nerve
2 Deep fibular nerve
3 Superficial fibular nerve

The **common fibular nerve** divides into the **deep fibular nerve** and the **superficial fibular nerve**. The **superficial fibular nerve** innervates the **fibularis longus** and **fibularis brevis** muscles. The **deep fibular nerve** innervates the **tibialis anterior, extensor digitorum longus, extensor hallucis longus, fibularis tertius, extensor hallucis brevis,** and **extensor digitorum brevis** muscles. Together these muscles extend the toes and dorsiflex the foot. It also innervates the skin on the anterior lower leg and the dorsal surface of the foot. "**Foot drop**" is a condition caused by severe damage to or paralysis of the common fibular nerve. The tibialis anterior muscle is paralyzed and the foot drops into plantar flexion. The individual has a high stepping gait so that the toes do not hit the ground while walking. The foot drops suddenly and makes a characteristic clopping sound.

Tibial Nerve
(tib•ee•al)

1 Tibial nerve

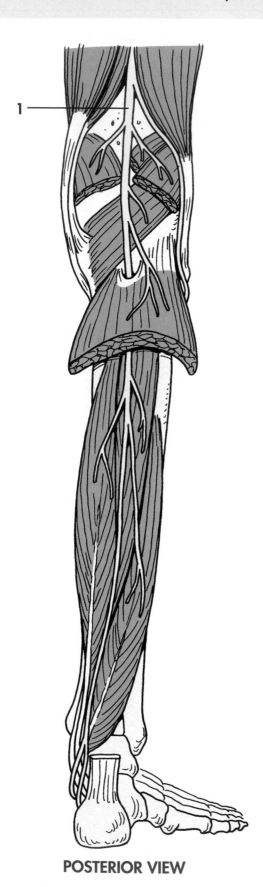

The **tibial nerve** innervates the muscles of the posterior compartment of the leg including the **plantaris, popliteus, gastrocnemius, soleus, flexor digitorum longus, flexor hallucis longus,** and **tibialis posterior.** One of its branches, the **medial plantar nerve,** innervates the **flexor digitorum brevis, abductor hallucis, flexor hallucis brevis,** and **first lumbrical.** The other branch, the **lateral plantar nerve,** innervates the **adductor hallucis, quadratus plantae, abductor digiti minimi, flexor digiti minimi, plantar interosseous, dorsal interosseous,** and **lumbricals** (lateral 3). It also innervates the skin of the dorsal surface of the lower leg and the plantar surface of the foot.

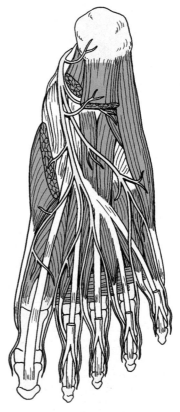

POSTERIOR VIEW

PLANTAR VIEW

Cranial Nerve V—Trigeminal Nerve

(try•**jem**•ih•nal)

Sensory Distribution

1 Superior orbital fissure
2 Trigeminal nerve
  2a Ophthalmic division (V_1)
  2b Maxillary division (V_2)
  2c Mandibular division (V_3)
3 Semilunar (gasserian) ganglion
4 Foramen ovale
5 Foramen rotundum
6 Mandibular foramen
7 Lingual nerve
8 Inferior alveolar nerve

Motor Distribution

1 Mandibular division (V_3)
2 Temporalis muscle
3 Medial pterygoid muscle
4 Lateral pterygoid muscle
5 Masseter muscle
6 Anterior belly of digastric muscle

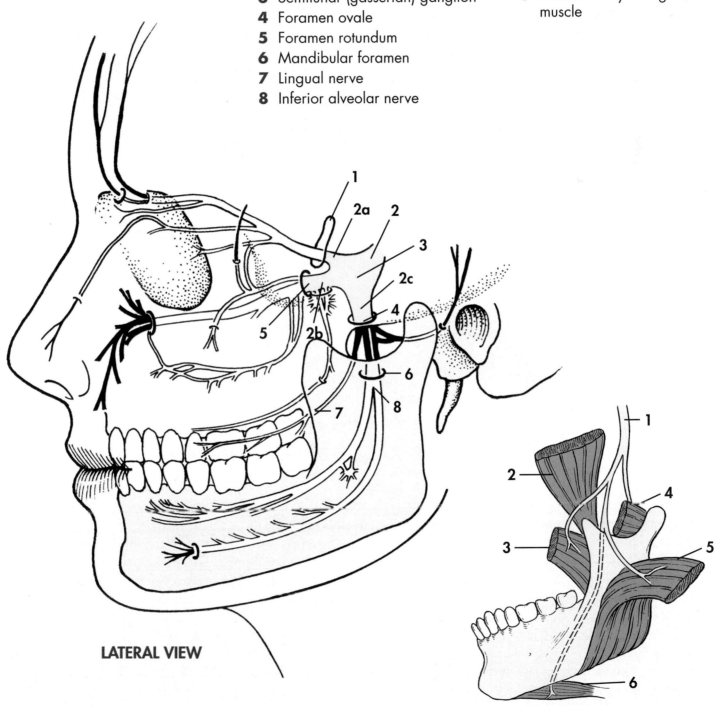

LATERAL VIEW

Cranial Nerve VII—Facial Nerve
(kray•ne•al)   (fay•shel)

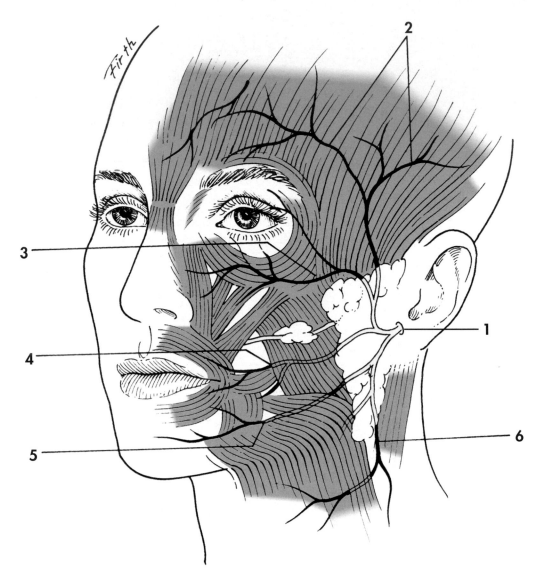

ANTERIOR LATERAL VIEW

1 Stylomastoid foramen
2 Temporal branches
3 Zygomatic branches
4 Buccal branches
5 Mandibular branches
6 Cervical branches

The **facial nerve** leaves the pons of the brain, enters the **temporal bone** through the **internal acoustic meatus,** and then emerges from that bone through the **stylomastoid foramen.** Its branches then radiate onto the lateral side of the face. The distribution of the five branches that innervate facial muscles can be visualized by placing five spread fingers on the side of the face: little finger above the eye (**temporal branch**), ring finger across the cheek (**zygomatic branch**), middle finger across upper jaw (**buccal branch**), index finger across lower jaw (**mandibular branch**), and thumb along lower neck (**cervical branch**).

The loss of facial muscle tone associated with **Bell's Palsy** is due to inflammation and swelling of the facial nerve thought to be caused by herpes simplex virus. The condition generally develops quickly and similarly may disappear fairly rapidly without treatment. The loss of muscle "tone" on one side of the face is generally obvious and can be seen in the drooping eyelids and sagging of the corner of the mouth.

Cranial Nerve XI—Spinoaccessory Nerve

(**kray**•nee•al)   (spy•no•ak•**sess**•o•ree)

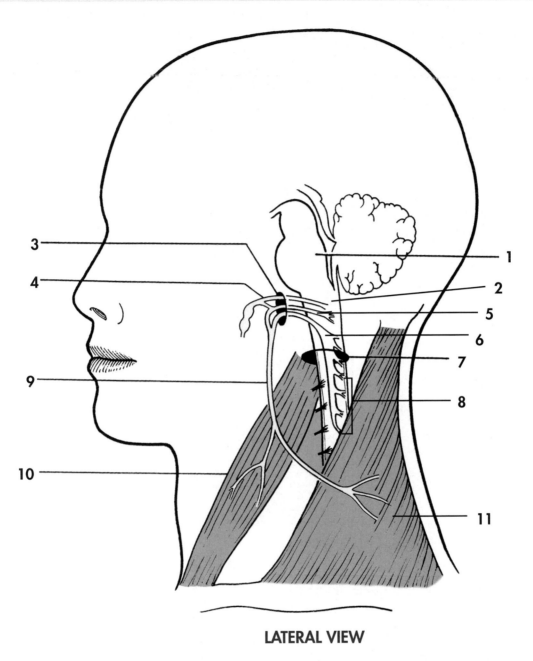

LATERAL VIEW

1 Pons
2 Medulla oblongata
3 Jugular foramen
4 Vagus nerve (X)
5 Cranial root

6 Spinal root
7 Foramen magnum
8 Cervical region of spinal cord (C_1–C_5)
9 Accessory nerve (XI)

10 Sternocleidomastoid muscle
11 Trapezius muscle

The (**spino**)**accessory nerve** is unique in that it is formed by both cranial and spinal components that combine, then diverge, the cranial portion joining the **vagus nerve**, the spinal portion descending to innnervate the **sternocleidomastoid** and **trapezius muscles**.

Cranial Nerves and Skull Passageways

(**kray**•ne•al)

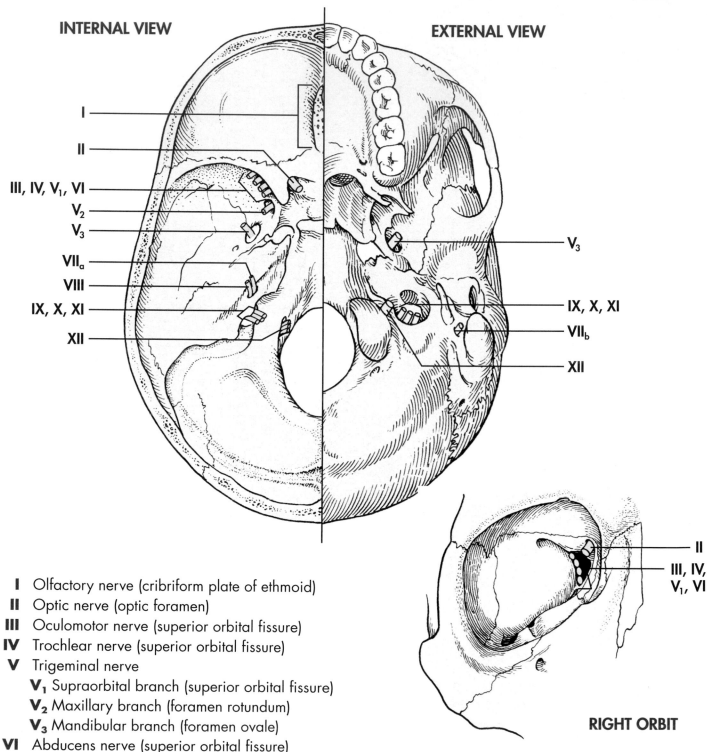

INTERNAL VIEW

I
II
III, IV, V₁, VI
V₂
V₃
VIIₐ
VIII
IX, X, XI
XII

EXTERNAL VIEW

V₃
IX, X, XI
VIIᵦ
XII

II
III, IV, V₁, VI

RIGHT ORBIT

I Olfactory nerve (cribriform plate of ethmoid)
II Optic nerve (optic foramen)
III Oculomotor nerve (superior orbital fissure)
IV Trochlear nerve (superior orbital fissure)
V Trigeminal nerve
  V₁ Supraorbital branch (superior orbital fissure)
  V₂ Maxillary branch (foramen rotundum)
  V₃ Mandibular branch (foramen ovale)
VI Abducens nerve (superior orbital fissure)
VII Facial nerve
  VIIₐ (internal acoustic meatus)
  VIIᵦ (stylomastoid foramen)
VIII Vestibulocochlaear/statoacoustic nerve (internal acoustic meatus)

IX Glossopharyngeal nerve (jugular foramen)
X Vagus nerve (jugular foramen)
XI Spinoaccessory (jugular foramen)
XII Hypoglossal (hypoglossal canal)

An Illustrated Atlas of the Skeletal Muscles

Sensory Innervation of Dermatomes of the Upper Limb
(der•ma•tomes)

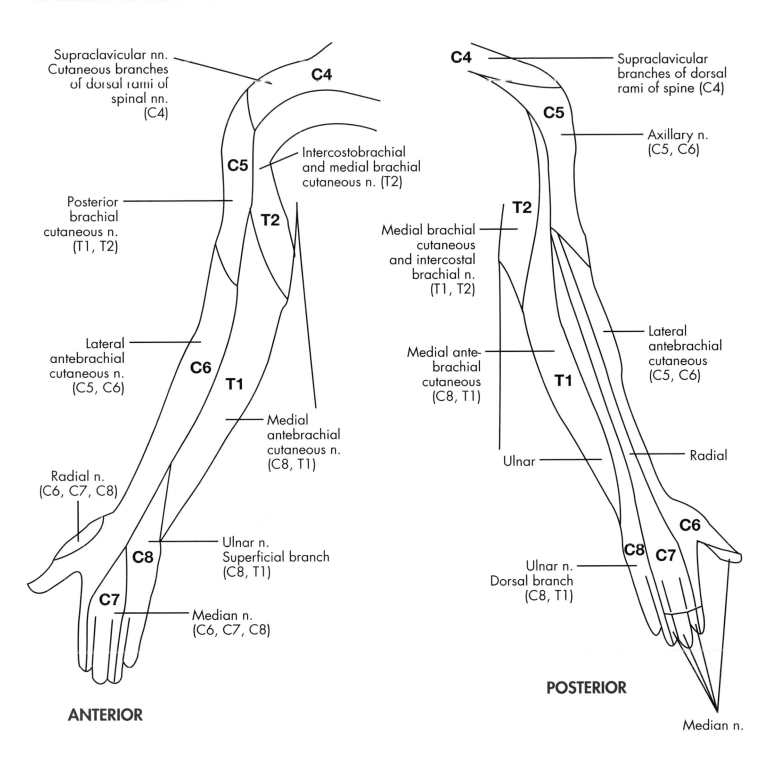

Supraclavicular nn.
Cutaneous branches
of dorsal rami of
spinal nn.
(C4)

C4

Supraclavicular
branches of dorsal
rami of spine (C4)

C5

Axillary n.
(C5, C6)

C5

Intercostobrachial
and medial brachial
cutaneous n. (T2)

Posterior
brachial
cutaneous n.
(T1, T2)

T2

T2

Medial brachial
cutaneous
and intercostal
brachial n.
(T1, T2)

Lateral
antebrachial
cutaneous n.
(C5, C6)

C6

T1

Medial ante-
brachial
cutaneous
(C8, T1)

T1

Lateral
antebrachial
cutaneous
(C5, C6)

Medial
antebrachial
cutaneous n.
(C8, T1)

Ulnar

Radial

Radial n.
(C6, C7, C8)

C8

Ulnar n.
Superficial branch
(C8, T1)

C8

C7

C6

C7

Ulnar n.
Dorsal branch
(C8, T1)

Median n.
(C6, C7, C8)

POSTERIOR

Median n.

ANTERIOR

M

Maisonneuve fracture a fracture of the proximal third of the fibula usually due to extreme external rotation of the foot

Malleolus a large and hammerhead-like process at the distal end of both the tibia and the fibula

Mallet finger a condition caused by hyperflexion of the distal phalanx in which the terminal phalanx cannot be extended

Meatus a canal or opening

Medial toward the midline of the body

Menisci medial and lateral fibrocartilaginous pads on the proximal surfaces of the tibia that act as shock absorbers in the knee joint; often torn, requiring knee surgery

Metacarpophalangeal joint joint between the metacarpals and phalanges

Metaphysis a section of bone between the epiphysis and the central portion of the shaft of long bones

Monteggia's fracture a fracture of the proximal ulna with dislocation of the head of the radius

Myography a graphic representation of the electrical changes during muscle contraction

N

Nasal refers to the nose

Non-displaced fracture a fracture in which the fragments of the bone remain in anatomical alignment

Notch refers to a recess or depression in a bone that might serve as an articulating surface or for the passage of blood vessels and/or nerves

O

Oblique fracture a fracture in which the break occurs across the bone at an angle to the long axis of the bone

Opposition movement of the thumb to approach or touch one or more of the fingertips

Oral refers to the mouth and associated structures

Orbital refers to the cup-shaped depression within which the eye is located, and associated structures

Origin attachment end of a muscle that remains relatively fixed during contraction

Osgood Schlatter Disease an apophyseal injury in which the tibial tuberosity may be partially or completely separated from the underlying growth plate due to excessive contraction of the quadriceps muscles

P

Palpation method of examining the body with the hands; it is particularly useful for locating skeletal and muscular landmarks and for diagnostic purposes, treatment such as massage, or identifying appropriate locations for injections

Pectoral refers to the chest region and associated structures

Pelvic refers to the hip region of the lower abdomen and associated structures

Perineal refers to the muscles of the floor of the abdominal region and the anus, vulva (female), and posterior scrotum (male)

Phalanx one of the bones of the fingers and toes (plural: phalanges)

Pivot (rotational) joints joint in which the rounded end of one bone fits into a ring formed by a depression on an adjoining bone and an encircling (annular) ligament; the bone turns around its long axis; uniaxial

Plantar refers to the inferior surface or sole of the foot

Plantar fasciitis inflammation of the plantar fascia

Plantar flexion movement of the foot that flexes the toes downward toward the sole

Plexus a network or interjoining of nerves or of blood vessels

Poland's syndrome a condition typically characterized by absence of pectoral muscles on the right side, combined with abnormalities in skeletal structure and skin, fusion of the digits (syndactyly) and assorted shoulder and thoracic anomalies

Pollicis refers to the thumb

Popliteal refers to the posterior area of the knee

Posterior the back (or dorsal) surface of the body or of an organ within the body

Pott's fracture fracturing of the lower part of the fibula and of the malleolus of the tibia as a result of extreme eversion of the foot

Process refers to any marked body prominence or projection usually serving for muscle and ligament attachment

Pronation medial rotation of the forearm causing the palm of the hand to face posteriorly

Protraction movement of an anatomical part forward; for example, protraction of the jaw

Proximal toward the attached end of a limb, or near the origin of a structure

R

Raphe the seamlike union of the two lateral halves of a part or organ

Rectus describes something that is straight

Referred pain pattern surface area, other than the immediate area of the source of the pain, in which pain is felt

Retraction movement of an anatomical part backward; for example, the retraction of the jaw

Rotation motion in which a bone turns around its own longitudinal axis

Rotator cuff four muscles (subscapularis, supraspinatus, infraspinatus, and teres minor) and their tendons that blend with the shoulder joint capsule strengthening this joint; subject to stretching or tearing due to extreme and prolonged movement of the arm

S

Sacral refers to the sacrum, the vertebral region located between the hip bones

Sacroiliac joint joint between the sacrum and adjacent surface of the ilium that can become dislocated and cause lower back pain by pressing on the sciatic nerve

Saddle joint the only saddle joint is between the trapezium carpal bone and metacarpal of the thumb; the concave-convex shape of both articulating surfaces allows a greater range of motion than permitted by the carpo-metacarpal joints of the remaining digits of the hand

Salter-Harris classification system of fractures a system of categorizing adolescent bone fractures on the basis of damage to one or more of the epiphysis, epiphyseal growth plate or the metaphysis

Scheuermanns's disease a degeneration of adjacent vertebral bodies causing narrowing of intervertebral disk spaces and potential protrusion of disks into the vertebral bodies causing decreased spinal height and increased spinal curvature

Sciatica pain in the lower back and hip radiating down the back of thigh into the leg, due to sciatic nerve dysfunction or due to a herniated lumbar disk pressing a nerve root

Scoliosis lateral curvature of one or more vertebral regions caused by unequal development of vertebral muscles or of both sides of the vertebrae

Sever's disease also known as "calcaneal apophysitis," is a fracture in which the calcaneal tuberosity is partially or completely separated from the underlying growth plate due to excessive contraction of the gastrocnemius muscle

Simple (closed) fracture a fracture in which the broken bone does not break through the skin

Slipped disc an intervertebral disk protruding out to one side, often producing pain and muscular weakness by pressing on a spinal nerve; also referred to as a herniated disk

Smith's fracture a fracture of the distal radius with displacement of the fragment toward the palmar surface

Sphincter a muscle (generally circular) surrounding an opening, whose contraction and relaxation serve to open and close that passageway

Spina bifida a congenital cleft of the vertebrae with hernial protrusion of the meningeal membranes and sometimes of the spinal cord

Spine a sharp, generally slender projection

Spiral fracture a fracture in which the fracture line "spirals around" the bone due to excessive twisting of the bone

"Spring" ligament the calcaneonavicular ligament that contributes to the formation of the arch of the foot

Sternal puncture a procedure in which a needle is inserted through the sternal surface of the thoracic cage to aspirate a sample of red bone marrow

Stress fractures fractures that occur in the navicular bone and metatarsals of the foot due to repetitive trauma

Styloid an elongate and tapered bony process for muscle attachment

Superficial refers to a position above another structure, generally from beneath the body surface outward

Superior above another structure; toward the head

Supination lateral rotation of the forearm causing the palm of the hand to face anteriorly

Suture an immovable, fibrous joint; all bones of the skull (except jaw joint) are united by sutures

Symphysis a joint in which bones are connected by fibro-cartilage, such as the intervertebral discs and the pubic symphysis

Synchondrosis an articulation in which bones are joined by hyaline cartilage

Syndactyly fusion of two or more of the digits of the hand

Syndesmosis a fibrous articulation permitting some degree of movement between two bones

Synergist a muscle that acts together with another muscle such as the biceps brachii and brachialis both causing flexion of the lower arm

Synovial joint a freely movable joint with a fluid-filled cavity surrounded by a joint capsule

T

Tarsal refers to the ankle

Tendon a band or sheet of dense connective tissue of variable thickness and width that attaches muscles to bones

Tennis elbow an inflammatory condition resulting from repetitive pronation-supination of the forearm, producing inflammation and degeneration of the bony attachment of the common extensor tendon and inflammation of the lateral epicondyle of the humerus; also known as "elbow tendinitis," the condition is aggravated by any action that places tension on the common extensor tendon, such as grasping a tennis racket, a golf club, a screwdriver, or a snow shovel

Thoracic refers to the chest region

TMJ syndrome pain around the temporomandibular joint caused by misalignment of teeth, trauma to the jaw, or arthritis

Transverse fracture a fracture in which the break occurs at a right angle to the long axis of the bone

Trigger point a hyperirritable site within a muscle and/or its tendon and other closely associated connective tissue; compression of a trigger point results in distribution of pain in an area (a referred pain pattern) usually somewhat distant from the location of the trigger point

Trochanter a large process for muscle attachment; the greater and lesser trochanters on the proximal end of the femur

Trochlear a structure that is shaped like a pulley

Tubercle a small, rounded process

Tuberosity a large roughened process on a bone, larger than a tubercle

U

Urethra tube-like passage that carries urine from the bladder

Uvula a hanging fleshy lobe in the middle of the posterior part of the soft palate

V

Vagina a canal in the female that leads from the external opening of the vestibule to the uterus

Ventral *See* anterior

Volar referring to the palm of the hand

W

Whiplash injury resulting from severe hyperextension of the neck

Wry neck (torticollis) a twisting of the neck to one side

Index